DO THE MATH WORKBOO

INTERMEDIATE ALGEBRA
THIRD EDITION

Michael Sullivan, III
Joliet Junior College

Katherine Struve
Columbus State Community College

Janet Mazzarella
Southwestern College

PEARSON

Boston   Columbus   Indianapolis   New York   San Francisco   Upper Saddle River
Amsterdam   Cape Town   Dubai   London   Madrid   Milan   Munich   Paris   Montreal   Toronto
Delhi   Mexico City   São Paulo   Sydney   Hong Kong   Seoul   Singapore   Taipei   Tokyo

Five-Minute Warm-Up R.2
Sets and Classification of Numbers

Recognizing symbols and notation. Fill each blank with the correct symbol.

1. subset _____   **2.** proper subset _____   **3.** null set _____

4. element of a set _____   **5.** not an element of a set _____   **6.** strictly greater than _____

7. strictly less than _____   **8.** greater than or equal to _____   **9.** less than or equal to _____

Every real number must fit into one of the three following categories: the number is <u>positive</u>, the number is <u>negative</u>, or the number is equal to <u>zero</u>. If x is any real number, use symbols to state each of the following:

10. x is positive _____   **11.** x is negative _____

Classifying numbers in the real number system. Use the roster method to list some elements of each of the following sets of numbers.

12. $\mathbb{N}$ natural numbers _____

13. W: whole numbers _____

14. $\mathbb{Z}$: integers _____

15. $\mathbb{Q}$: rational numbers _____

16. irrational numbers _____

17. $\mathbb{R}$: real numbers _____

18. Find the decimal representation of the rational number $\dfrac{1}{9}$ and then state whether this is a terminating decimal or a nonterminating (repeating) decimal.

$$\frac{1}{9} = \underline{\hspace{2cm}} \qquad \underline{\hspace{3cm}}$$

19. Find the decimal representation of the rational number $\dfrac{3}{8}$ and then state whether this is a terminating decimal or a nonterminating (repeating) decimal.

$$\frac{3}{8} = \underline{\hspace{2cm}} \qquad \underline{\hspace{3cm}}$$

Guided Practice R.2
Sets and Classification of Numbers

Objective 1: Use Set Notation

1. Explain the difference between a subset and a proper subset.

1. _____

2. Is the empty set a subset of every set?   2. _____

3. Is every set a subset of itself?   3. _____

4. If set A is a subset of set B $\left(\text{written } A \subseteq B\right)$, what is the necessary condition for A to be a proper subset of B?

4. _____

5. Let $D = \{a, b, c\}$. List every possible subset of D. (*Hint: there should be 8 subsets.*)

5. _____

6. Let $X = \{-2, -1, 0, 1, 2\}$, $Y = \{x | x \text{ is a natural number}\}$, $Z = \{1, 3, 5, 7\}$, and $W = \{-1, 0, 1\}$. Write TRUE or FALSE to each statement. *See Example 3 and Example 4*

(a) $W \subset X$ _____

(b) $W \subseteq X$ _____

(c) $X \subset W$ _____

(d) $Z \subset Y$ _____

(e) $\emptyset \subset Y$ _____

(f) $\{-1, 0, 1\} \subset W$ _____

(g) $\{-1, 0, 1\} \subseteq W$ _____

(h) $X \subset Y$ _____

(i) $0 \in Y$ _____

(j) $\frac{2}{3} \notin Y$ _____

(k) Every element of W is also an element of X.

_____

(l) $\frac{1}{2} \in Y$ _____

Objective 2: Classify Numbers

7. Let $S = \left\{\dfrac{4}{0}, -1, 1.\overline{3}, \pi, \dfrac{-6}{3}, 100{,}000, -\dfrac{5}{4}, \sqrt{9}, \dfrac{0}{2}, \sqrt{5}, 0.25, \sqrt{-16}\right\}$. *See Example 5*

(a) Rewrite S, simplifying elements when possible _____

List the elements of S that are:

(b) Natural numbers _____

(c) Whole numbers _____

(d) Integers _____

(e) Rational numbers _____

(f) Irrational numbers _____

(g) Real numbers _____

Do the Math Exercises R.2
Sets and Classification of Numbers

In Problems 1 and 2, write each set using the roster method.

1. $\{x \mid x$ is an integer between -4 and $6\}$ **2.** $\{x \mid x$ is a whole number less than $0\}$ 1. _____

2. _____

In Problems 3 – 6, let A = {1, 3, 5, 7, 9}, B = {2, 4, 6, 8}, C = {1, 2, 3, 4, 5, 6, 7, 8, 9}, and D = {8, 6, 4, 2}. Write TRUE or FALSE to each statement. Be sure to justify your answer.

3. $A \subseteq C$ **4.** $A \subset C$ 3. _____

4. _____

5. $B = D$ **6.** $\varnothing \subset B$ 5. _____

6. _____

In Problems 7 and 8, fill in the blank with the appropriate symbol, $\in$ or $\notin$.

7. $4.\overline{5}$ ____ $\{x \mid x$ is a rational number$\}$ **8.** 0 ____ $\{x \mid x$ is a natural number$\}$ 7. _____

8. _____

In Problems 9 and 10, list numbers in each set that are (a) Natural numbers, (b) Integers, (c) Rational numbers, (d) Irrational numbers, (e) Real numbers.

	9. $\left\{-5, 4, \dfrac{4}{3}, -\dfrac{7}{5}, 5.\overline{1}, \pi\right\}$	**10.** $\left\{13, 0, -4.5656..., 2.43, \sqrt{2}, \dfrac{8}{7}, \sqrt{-9}\right\}$
(a)		
(b)		
(c)		
(d)		
(e)		

In Problem 11, approximate the number by (a) truncating and (b) rounding to the indicated number of decimal places.

11. -9.9999; 2 decimal places   11a. _____

11b. _____

12. On the real number line, label the points

with coordinates: $4, 2.5, -\dfrac{5}{3}, -0.5$ ⟶

In Problems 13 and 14, replace the question mark by <, >, or =, whichever is correct.

13. $\dfrac{2}{3}$? $\dfrac{2}{5}$

14. $\dfrac{1}{3}$? 0.3

13. _____

14. _____

15. Cisco Systems Cisco systems stock lost $0.37 in a recent trading day. Express this loss as a rational number. (SOURCE: *Yahoo! Finance*)

15. _____

16. Use your calculator to express $\dfrac{8}{7}$ rounded to three decimal places.

16. _____

17. Use your calculator to express $\dfrac{8}{7}$ truncated to three decimal places.

17. _____

18. Express $\dfrac{2}{3}$ as a decimal. Is this decimal terminating or repeating?

18. _____

_____

19. Express $-\dfrac{4}{25}$ as a decimal. Is this decimal terminating or repeating?

19. _____

_____

20. (a) Are there any real numbers that are both rational and irrational? **(b)** Are there any real numbers that are neither? **(c)** Explain your reasoning.

20a. _____

20 b. _____

20c. _____

Five-Minute Warm-Up R.3
Operations on Signed Numbers; Properties of Real Numbers

Definitions and vocabulary. In your own words, write a brief definition for each of the following.

1. absolute value _____

2. product _____

3. quotient _____

4. sum _____

5. difference _____

Division Properties For any nonzero real number a,

6. $\dfrac{0}{a} =$ _____   **7.** $\dfrac{a}{a} =$ _____   **8.** $\dfrac{a}{0} =$ _____

In Problems 9 and 10, identify the (a) divisor; (b) dividend; and (c) quotient.

9. $4\overline{)12}$ **(a)** _____; **(b)** _____; **(c)** _____

10. $\dfrac{5}{20}$ **(a)** _____; **(b)** _____; **(c)** _____

In Problems 11 – 14, perform the indicated operation. Express all rational numbers in lowest terms.

11. $\dfrac{3}{4} + \dfrac{1}{2}$ 11. _____

12. $\dfrac{9}{5} - \dfrac{4}{5}$ 12. _____

13. $\dfrac{5}{8} \cdot \dfrac{12}{25}$ 13. _____

14. $\dfrac{25}{12} \div \dfrac{30}{27}$ 14. _____

Guided Practice R.3
Operations on Signed Numbers; Properties of Real Numbers

● **Objective 2: Add and Subtract Signed Numbers**

1. To add two real numbers that are the same sign, add their absolute values.

    (a) If both numbers are positive, the sign of the sum is _____ .   1a. _____

    (b) If both numbers are negative, the sign of the sum is _____ .   1b. _____

2. To add two real numbers that are different signs, subtract the absolute value of the smaller number from the absolute value of the larger number. The sign of the sum is _____ .   2. _____

Objective 3: Multiply and Divide Signed Numbers

3. To multiply (or divide) two real numbers, we multiply (or divide) the numbers and determine the sign of the product (or quotient) according to the following rule. Complete the chart, where + indicates that the sign of the number is positive and − indicates the sign of the number is negative.

$\bullet$ / $\div$	+	−
+		
−		

● **Objective 4: Perform Operations on Fractions** *(See textbook Example 10)*

4. Find the least common denominator of the rational numbers $\dfrac{7}{12}$ and $\dfrac{5}{18}$. Then rewrite each rational number with the least common denominator.

Step 1: Factor each denominator.

    List the prime factors of 12:   **(a)** _____

    List the prime factors of 18:   **(b)** _____

Step 2: Write down the common factor(s) between each denominator. Then copy the remaining factors.

    List the common factors: **(c)** _____

    List all remaining factors: **(d)** _____

Step 3: Multiply the factors listed in Step 2. The product is the least common denominator.

    Multiply the factors listed in (c) and (d):   **(e)** _____

Step 4: Rewrite $\dfrac{7}{12}$ and $\dfrac{5}{18}$ with a denominator of 36.

    To rewrite $\dfrac{7}{12}$ with a denominator of 36, we multiply the fraction by $\dfrac{3}{3}$ because $a \cdot 1 = a$.

    (f) $\dfrac{7}{12} \cdot \dfrac{3}{3} = \dfrac{\quad}{36}$

(g) $\dfrac{5}{18} \cdot \dfrac{\rule{1cm}{0.4pt}}{\rule{1cm}{0.4pt}} = \dfrac{\rule{1cm}{0.4pt}}{36}$

Multiply to find a fraction
equivalent to $\dfrac{5}{18}$ whose
denominator is 36.

Objective 5: Know the Associative and Distributive Properties

5. Write an example for each of the properties of the real number system.

(a) Identity Property of Addition   **(a)** _____

(b) Identity Property of Multiplication   **(b)** _____

(c) Inverse Property of Addition   **(c)** _____

(d) Inverse Property of Multiplication   **(d)** _____

(e) Double Negative Property   **(e)** _____

(f) Commutative Property of Addition   **(f)** _____

(g) Commutative Property of Multiplication   **(g)** _____

(h) Multiplication Property of Zero   **(h)** _____

(i) Associative Property of Addition   **(i)** _____

(j) Associative Property of Multiplication   **(j)** _____

(k) Distributive Property   **(k)** _____

6. Use the Distributive Property to remove the parentheses. *See textbook Example 13*

(a) $3(2x - 5)$ **(b)** $-12(3y - 1)$

6a. _____

6b. _____

(c) $(5n + 2) \cdot 4$ **(d)** $\dfrac{2}{3}(6x - 24)$

6c. _____

6d. _____

Do the Math Exercises R.3
Operations on Signed Numbers; Properties of Real Numbers

In Problems 1 and 2, determine (a) the additive inverse and (b) the multiplicative inverse of the given number.

1. $-\dfrac{1}{5}$

2. 10

1a. _____

1b. _____

2a. _____

2b. _____

3. Use the Distributive Property to remove the parentheses: $-\dfrac{2}{3}(3x + 15)$

3. _____

4. Use the Reduction Property to simplify: $\dfrac{40}{16}$

4. _____

In Problems 5 – 16, perform the indicated operation. Express all rational numbers in lowest terms.

5. $|-4| + 12$

6. $7 \cdot (-15)$

5. _____

6. _____

7. $-\dfrac{7}{3} \cdot \left(-\dfrac{12}{35}\right)$

8. $-\dfrac{10}{3} \div \dfrac{15}{21}$

7. _____

8. _____

9. $\dfrac{\dfrac{18}{7}}{\dfrac{3}{14}}$

10. $\dfrac{8}{5} - \dfrac{18}{5}$

9. _____

10. _____

11. $-\dfrac{17}{45} - \dfrac{23}{24}$

12. $\dfrac{2}{3} \div 8$

11. _____

12. _____

13. $\left|-5.4+10.5\right|$

14. $\dfrac{20}{\dfrac{5}{4}}$

13. _____

14. _____

15. $-\dfrac{5}{2}-4$

16. $-\dfrac{5}{6}-\dfrac{1}{4}+\dfrac{5}{24}$

15. _____

16. _____

In Problems 17 – 20, state the property that is being illustrated.

17. $3+(4+5)=(3+4)+5$

18. $5\cdot\dfrac{1}{5}=1$

17. _____

18. _____

19. $5\cdot\dfrac{4}{4}=5$

20. $(x+2)\cdot3=3\cdot(x+2)$

19. _____

20. _____

21. Peaks and Valleys In Louisiana, the highest elevation is Driskill Mountain (535 feet above sea level); the lowest elevation is New Orleans (8 feet below sea level). What is the difference between the highest and lowest elevation?

21. _____

22. Why Is the Product of Two Negatives Positive? In this problem we use the Distributive Property to illustrate why the product of two negative real numbers is positive.

(a) Express the product of any real number a and 0.

22a. _____

(b) Use the Additive Inverse Property to write 0 from part (a) as $b+(-b)$.

22b. _____

(c) Use the Distributive Property to distribute the a into the expression in part (b).

22c. _____

(d) Suppose that $a<0$ and $b>0$. What can be said about the product ab? Now, what must be true regarding the product $a(-b)$ in order for the sum to be zero?

22d. _____

Five-Minute Warm-Up R.4
Order of Operations

Definitions and vocabulary. Fill in the blank for each of the following.

1. Integer exponents provide a shorthand device for representing repeated multiplications of a real number. In the expression 4^3,

    (a) 4 is called the _____

    1a. _____

    (b) 3 is called the _____

    1b. _____

    (c) To evaluate the expression, multiply _____ = _____.

    1c. _____

2. Consider the expression $3x$.

    (a) Another word for raising the expression $3x$ to the second power is to say $3x$ _____.

    2a. _____

    (b) This is written _____.

    2b. _____

3. Consider the expression $y + 1$.

    (a) Another word for raising the expression $y + 1$ to the third power is to say $y + 1$ _____.

    3a. _____

    (b) This is written _____.

    3b. _____

4. Identify the base for each expression.

    (a) -5^2 _____   **(b)** $(-2x)^4$ _____

    4a. _____

    4b. _____

In Problems 5 – 6, perform the indicated operation. Express all rational numbers in lowest terms.

5. $\left| \left(-\dfrac{3}{2} \right) \left(-\dfrac{3}{2} \right) \left(-\dfrac{3}{2} \right) \right|$

    5. _____

6. $-\dfrac{5}{6} + \left(-\dfrac{4}{9} \right)$

    6. _____

Guided Practice R.4
Order of Operations

Objective 1: Evaluate Real Numbers with Exponents

1. Evaluate each expression. (*See textbook Examples 1 and 2*)

(a) 6^2

(b) $(-6)^2$

(c) -6^2

1a. _____

1b. _____

1c. _____

(d) $-(-6)^2$

(e) $\left(-\dfrac{4}{5}\right)^3$

(f) $\left(-\dfrac{4}{5}\right)^4$

1d. _____

1e. _____

1f. _____

2. Raising a negative number to an even exponent results in a _____ number.

2. _____

3. Raising a negative number to an odd exponent results in a _____ number.

3. _____

Objective 2: Use the Order of Operations to Evaluate Expressions

4. Complete the table:

		Order of Operations	
(a)	P	_____	When an expression has multiple parentheses, begin with the innermost parentheses and work outward.
(b)	E	_____	Work from left to right.
(c)	M/D	_____	Work from left to right.
(d)	A/S	_____	Work from left to right.

5. Evaluate: $\dfrac{3 \cdot (-2)^2}{-3^2 - 6 \cdot 3}$ *(See textbook Example 7)*

Step 1:   Evaluate exponents.

(a) $\dfrac{3 \cdot (-2)^2}{-3^2 - 6 \cdot 3} =$

_____

Step 2:   Multiply.

(b)_____

Step 3:   Subtract.

(c) _____

Step 4:   Divide out the common factor.

(d) _____

Sullivan/Struve, *Intermediate Algebra*, 3e

Do the Math Exercises R.4
Order of Operations

In Problems 1 – 18, evaluate each expression.

1. $(-3)^2 + (-2)^2$

2. $-5 + 3 \cdot 12$

3. $-4^2 - (-4)^2$

4. $3\left[15 - (7 - 3)\right]$

5. $2 + 5(8 - 5)$

6. $\dfrac{5 - (-7)}{4}$

7. $\left|6 \cdot 2 - 5 \cdot 3\right|$

8. $\dfrac{12 - 4}{-2}$

9. $\dfrac{2 \cdot 4 - 5}{4^2 + (-2)^3} + \dfrac{3^2}{2^3}$

10. $3\left[2 + 3(1 + 5)\right]$

11. $\dfrac{\dfrac{4}{4^2 - 1} - \dfrac{3}{5(7 - 5)}}{\dfrac{-(-2)^2}{4 \cdot 7 + 2}}$

12. $(3^2 - 3)(3 - (-3)^3)$

1. _____

2. _____

3. _____

4. _____

5. _____

6. _____

7. _____

8. _____

9. _____

10. _____

11. _____

12. _____

13. $-2(5-2)-(-5)^2$

14. $-2\left(4+\left|2\cdot 3-5^2\right|\right)$

13. _____

14. _____

15. $\left|6(3\cdot 2-10)\right|$

16. $\left|4(2\cdot 5+(-3)4)\right|$

15. _____

16. _____

17. $\dfrac{3(5+2^2)}{2\cdot 3^3}$

18. $\left(\dfrac{2}{3}\right)^2\cdot\left(\dfrac{1+2^3}{2^3-2}\right)$

17. _____

18. _____

19. Geometry The surface area of a right circular cylinder whose radius is 6 inches and height is 10 inches is given approximately by $2\cdot\pi\cdot 6^2+2\cdot\pi\cdot 6\cdot 10$ in^2 where $\pi\approx 3.1416$. Evaluate this expression rounded to the nearest hundredth.

19. _____

20. Insert parentheses in order to make the statement true:

    (a) $3+5\cdot 6-3=18$

    (b) $3+5\cdot 6-3=24$

20a. _____

20b. _____

21. Explain why $\dfrac{2+7}{2+9}\neq\dfrac{7}{9}$.

Sullivan/Struve, *Intermediate Algebra*, 3e
Copyright © 2014 Pearson Education, Inc.

Five-Minute Warm-Up R.5
Algebraic Expressions

1. Use the Distributive Property to remove the parentheses.

(a) $-5(2x - 3)$

1a. _____

(b) $(4a + 3) \cdot 9$

1b. _____

(c) $\dfrac{2}{3}(6n - 15)$

1c. _____

(d) $-\dfrac{5}{4}\left(\dfrac{8}{5}z - \dfrac{24}{25}\right)$

1d. _____

2. Evaluate each expression.

(a) $12 - 4 \cdot 5$

2a. _____

(b) $\dfrac{2 - 6}{4 + 2(-5)}$

2b. _____

(c) $\dfrac{2}{3} \cdot 6^2 - 4 + 2^3$

2c. _____

(d) $\dfrac{3 \cdot (-2) + 5}{6} - \dfrac{-(-2)^2 + 8}{9}$

2d. _____

Guided Practice R.5
Algebraic Expressions

Objective 1: Translate English Expressions into Mathematical Language

1. Express each English phrase using an algebraic expression. *(See textbook Example 1)*

(a) The product of -2 and some number x **(b)** 6 less than half of a number y

(c) 8 divided into a number m **(d)** A number x subtracted from 25

1a. _____

1b. _____

1c. _____

1d. _____

Objective 2: Evaluate Algebraic Expressions

2. Evaluate the expression $-x^2 - 4x + 5$ for $x = -3$. *(See textbook Example 2)*

2. _____

Objective 3: Simplify Algebraic Expressions by Combining Like Terms

3. Simplify each expression by combining like terms. *(See textbook Examples 4 and 5)*

(a) $5v + v - 6v$ **(b)** $4x^2 + 2x^2$

3a. _____

3b. _____

4. Simplify each algebraic expression. *(See textbook Example 6 and 7)*

(a) $3(x - 2) + 5$ **(b)** $\dfrac{5}{3}(6x + 9) - 4x - \dfrac{1}{2}(8x - 2)$

4a. _____

4b. _____

Objective 4: Determine the Domain of a Variable

5. The set of values that a variable may assume is called the _____ of the variable.   5. _____

6. In the expression $\dfrac{1}{x}$, what value(s) for x must be excluded from the domain?   6. _____

7. In general, we exclude any values of the variable which causes _____

8. Determine which of the numbers are in the domain of the variable. *(See textbook Example 8)*

   (a) $\dfrac{12}{2x - 4}$ (b) $\dfrac{x - 3}{x + 5}$ 8a. _____

   $x = 0; \quad x = -2; \quad x = 2; \quad x = 4$ $x = 3; \quad x = 5; \quad x = -5; \quad x = 0$ 8b. _____

Sullivan/Struve, *Intermediate Algebra,* 3e
Copyright © 2014 Pearson Education, Inc.

Do the Math Exercises R.5
Algebraic Expressions

In Problems 1 – 4, evaluate each expression for the given value of the variable.

1. $y^2 - 4y + 5$; $y = 3$

2. $-2z^2 + z + 3$; $z = -4$

1. _____

2. _____

3. $|4z - 1|$; $z = -\dfrac{5}{2}$

4. $\dfrac{|3 - 5z|}{(z - 4)^2}$; $z = 4$

3. _____

4. _____

In Problems 5 – 14, simplify each expression by combining like terms.

5. $5y + 2y$

6. $8x - 9x + 1$

5. _____

6. _____

7. $\dfrac{3}{10}y + \dfrac{4}{15}y$

8. $-x - 3x^2 + 4x - x^2$

7. _____

8. _____

9. $10y + 3 - 2y + 6 + y$

10. $3(2y + 5) - 6(y + 2)$

9. _____

10. _____

11. $\dfrac{1}{2}(20x - 14) + \dfrac{1}{3}(6x + 9)$

12. $-4(w - 3) - (2w + 1)$

11. _____

12. _____

13. $\dfrac{4}{3}(5y + 1) - \dfrac{2}{5}(3y - 4)$

14. $0.4(2.9x - 1.6) - 2.7(0.3x + 6.2)$

13. _____

14. _____

In Problems 15 – 18, express each English phrase using an algebraic expression.

15. The difference of 10 and a number y

16. The ratio of a number x and 5

15. _____

16. _____

17. A number z increased by 30

18. Twice some number x decreased by the ratio of a number y and 3

17. _____

18. _____

19. Which of the following are in the domain of the variable? $\dfrac{x+1}{x^2+5x+4}$

19. _____

(a) $x = 5$ (b) $x = -1$ (c) $x = -4$ (d) $x = 0$

20. Area of a Triangle The algebraic expression $\dfrac{1}{2}(h+2)h$ represents the area of a triangle whose base is two centimeters longer than its height h. See the figure. Evaluate the algebraic expression for $h = 10$ centimeters.

20. _____

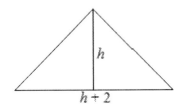

21. Write an English phrase that would translate into the given mathematical expression.

(a) $2x - 3$ _____

(b) $2(x - 3)$ _____

Five-Minute Warm-Up 1.1
Linear Equations in One Variable

1. Determine the additive inverse of $-\dfrac{1}{2}$.

   1._____

2. Determine the multiplicative inverse of 6.

   2._____

3. Use the Reduction Property to simplify the expression: $\dfrac{2}{3} \cdot \dfrac{3}{2}x$

   3._____

4. Find the Least Common Denominator of $\dfrac{5}{12}$ and $\dfrac{7}{15}$.

   4._____

5. Use the Distributive Property to remove the parenthesis: $-3(2x-5)$

   5._____

6. What is the coefficient of $-x$?

   6._____

7. Simplify by combining like terms: $-2(3x-1)+5+7x$

   7._____

8. Simplify: $6\left(\dfrac{x}{3}+\dfrac{5x}{6}\right)$

   8._____

9. Evaluate the expression $\dfrac{2}{5}(5x-10)+3x$ when $x=-3$.

   9._____

10. Is $x=-1$ in the domain of $\dfrac{6}{1-x}$?

   10._____

Guided Practice 1.1
Linear Equations in One Variable

Objective 1: Determine Whether a Number Is a Solution to an Equation

1. How do you determine if a value for a variable is a solution to an equation?

Objective 2: Solve Linear Equations *(See textbook Example 5)*

2. The goal in solving any linear equation is to get the variable by itself with a coefficient of 1, that is, to isolate the variable. List 5 steps you might use when solving a linear equation in one variable.

a. _____

b. _____

c. _____

d. _____

e. _____

3. Solve the linear equation $\dfrac{2x+3}{2} - \dfrac{x-1}{4} = \dfrac{x+5}{12}$.

Step 1: Remove all parentheses using the Distributive Property.

Determine the LCD: **(a)** _____

Use the Multiplication Property of Equality: **(b)** ___ $\cdot \left(\dfrac{2x+3}{2} - \dfrac{x-1}{4} \right) = \left(\dfrac{x+5}{12} \right) \cdot$ ___

Use the Distributive Property: **(c)** _____

Divide out the common factors: **(d)** _____

(e) Use _____ $12x + 18 - 3x + 3 = x + 5$

Step 2: Combine like terms on each side of the equation.

Combine like terms: **(f)** _____

Step 3: Use the Addition Property of Equality to get all variables on one side of the equation and all constants on the other side.	Subtract x from both sides:	**(g)** $9x + 21$ _____ $= x + 5$ _____
	Simplify:	**(h)** _____
	Subtract 21 from both sides:	**(i)** $8x + 21$ _____ $= 5$ _____
	Simplify:	**(j)** _____
Step 4: Use the Multiplication Property of Equality to get the coefficient on the variable to equal 1.	Divide both sides by 8:	**(k)** _____
Step 5: Check Verify the solution.	Substitute your value for x into the original equation:	$\dfrac{2x + 3}{2} - \dfrac{x - 1}{4} = \dfrac{x + 5}{12}$
	Simplify:	
	State the solution set:	**(l)** _____

Objective 3: Determine Whether an Equation Is a Conditional Equation, an Identity, or a Contradiction

4. Solve the linear equation $-2(3x - 1) = 4 - 6(x + 3)$. State whether the equation is an identity, contraction or conditional equation. *(See textbook Example 7)*

(a) Is the last statement of the solution true or false? _____

(b) Therefore the solution set is _____

(c) Is this equation an identity, contradiction or a conditional equation? _____

5. Solve the linear equation $3 - 5(2x + 1) - 9 = -4x + 5 - 2(3x + 8)$. State whether the equation is an identity, contraction or conditional equation. *(See textbook Example 8)*

(a) Is the last statement of the solution true or false? _____

(b) Therefore the solution set is _____

(c) Is this equation an identity, contradiction or a conditional equation? _____

Sullivan/Struve, *Intermediate Algebra*, 3e
Copyright © 2014 Pearson Education, Inc.

Guided Practice 1.2
An Introduction to Problem Solving

Objective 1: Translate English Sentences into Mathematical Statements

1. List the five categories of problems we will be modeling in this course. Give a brief description of each.

_____

_____

_____

_____

_____

2. List six steps for solving problems with mathematical models.

_____   _____

_____   _____

_____   _____

Objective 2: Model and Solve Direct Translation Problems

3. If n represents the first of three unknown consecutive integers, how would you represent the next two

integers? **(a)** _____   **(b)** _____

4. If x represents the first of four unknown consecutive even integers, how would you represent the next

three even integers? **(a)** _____   **(b)** _____   **(c)** _____

5. If p represents the first of four unknown consecutive odd integers, how would you represent the next three

odd integers? **(a)** _____   **(b)** _____   **(c)** _____

6. What is the formula for simple interest? Identify what each of the variables represent.

_____

Objective 3: Model and Solve Mixture Problems *(See textbook Example 8)*

7. Candy Store Valentine's Day is coming up so Andy decided to buy chocolates for his co-workers at $4.50 per pound and truffles for his girlfriend at $7.50 per pound. If he purchased a total of 11 pounds of candy and spent $58.50, how many pounds of each type did he buy?

(a) Eleven pounds of candy were purchased. If p represents the number of pounds of chocolates purchased and the rest of the candies were truffles, write an expression that represents the pounds of truffles. _____

(b) Complete the following table.

	Price $/Pound	Number of Pounds	Revenue
Chocolates			
Truffles			
Blend			

(c) Write and solve an equation that models Andy's purchase.

(d) Answer the question._____

Objective 4: Model and Solve Uniform Motion Problems *(See textbook Example 9)*

8. Boats Two boats are traveling toward the same port from opposite directions. They are 63 miles apart and one boat is traveling 6 mph faster than the other. If the boats both reach the port in 4 hours and 30 minutes, find the speed for each boat.

(a) What do you know about the distances travelled by the two boats? _____

(b) Complete the following table.

	Rate, mph	Time, hours	Distance, miles
Slower boat			
Faster boat			
Total			

(c) Write and solve an equation that models the distance travelled by the boats.

(d) Answer the question._____

Do the Math Exercises 1.2
An Introduction to Problem Solving

1. 50 is 90% of what?

2. 90 is what percent of 120?

1. _____

2. _____

In Problems 3 – 6, translate each of the following English statements into a mathematical statement. Then solve the equation.

3. The difference between 10 and a number z is 6.

4. The sum of two times y and 3 is 16.

3. _____

4. _____

5. Five times a number x is equivalent to the difference of three times x and 10.

6. 40% of a number equals the difference between the number and 10.

5. _____

6. _____

7. **Number Sense** Pattie is thinking of two numbers. She says that one of the numbers is 8 more than the other number and the sum of the numbers is 56. What are the numbers?

7. _____

8. **Consecutive Integers** The sum of four consecutive odd integers is 104. Find the integers.

8. _____

9. Investments Johnny is a shrewd 8-year old. For Christmas, his grandparents gave him   9. _____
$10,000. Johnny decides to invest some of the money in a savings account that pays 2%
per annum and the rest in a stock fund paying 10% per annum. Johnny wants his
investments to yield 7% per annum. How much should Johnny put into each account?

10. Coins Diana has been saving nickels and dimes. She opened up her piggy bank and   10. _____
determined that it contained 48 coins worth $4.50. Determine how many nickels and
dimes were in the piggy bank.

11. Antifreeze The cooling system of a car has a capacity of 15 liters. If the system is   11. _____
currently filled with a mixture that is 40% antifreeze, how much of this mixture should be
drained and replaced with pure antifreeze so that the system is filled with a solution that
is 50% antifreeze?

12. Candy "Sweet Tooth!" candy store sells chocolate-covered almonds for $6.50 per pound   12. _____
and chocolate-covered peanuts for $4.00 per pound. The manager decides to make a
bridge mix that combines the almonds with the peanuts. She wants the bridge mix to sell
for $5.00 per pound, and there should be no loss in revenue from selling the bridge mix
rather than the almonds and peanuts alone. How many pounds of chocolate-covered
almonds and chocolate-covered peanuts are required to create 50 pounds of bridge mix?

13. Cars Two cars leave from the same location an travel in opposite directions along a   13. _____
straight road. One car travels 30 miles per hour while the other travels at 45 miles per
hour. How long will it take the two cars to be 255 miles apart?

Sullivan/Struve, *Intermediate Algebra*, 3e
Copyright © 2014 Pearson Education, Inc.

Five-Minute Warm-Up 1.3
Using Formulas to Solve Problems

1. *List for the formulas for each of the following geometric figures.*

(a) Square	**(b) Rectangle**
Area:_____ Perimeter:_____	Area:_____ Perimeter:_____
(c) Triangle	**(d) Trapezoid**
Area:_____ Perimeter:_____	Area:_____ Perimeter:_____
(e) Parallelogram	**(f) Circle**
Area:_____ Perimeter:_____	Area:_____ Circumference:_____

In Problems 2 – 6, (a) round and then (b) truncate each decimal to the indicated number of places.

2a. _____

2. 15.96145; 3 decimal places

2b. _____

3a. _____

3. −0.098 ; 2 decimal places

3b. _____

4a. _____

4. 9.55; nearest whole number

4b. _____

5a. _____

5. 100.73; nearest tenth

5b. _____

6a. _____

6. $5.76715; nearest cent

6b. _____

Guided Practice 1.3
Using Formulas to Solve Problems

Objective 1: Solve for a Variable in a Formula

1. Explain the meaning of the expression "solve for the variable".

2. Solve the formula for the indicated variable. *(See textbook Examples 2 and 3)*

(a) $V = \dfrac{1}{3}Bh$ for h (b) $A = P + Prt$ for P

2a. _____

2b. _____

Objective 2: Use Formulas to Solve Problems *(See textbook Examples 4 and 5)*

3. The sum of the measures of the interior angles of a triangle is 180°. The measure of the first angle is 15° less than the second. The measure of the third angle is 45° more than half of the second. Find the measure of each interior angle of the triangle.

(a) **Step 1: Identify** What formula is needed to solve this problem? _____

(b) **Step 2: Name**
   If x represents the value of the second angle, what expression represents the value of the first angle?

   _____

   What expression represents the value of the third angle?   _____

(c) **Step 3: Translate** Substitute the values from Step 2 into the formula from Step 1.

$$x° + y° + z° = 180°$$

_____

(d) Step 4: Solve

(e) Step 5: Check Substitute the value you found for *x* into your equation. Does the sum of the angles equal 180° ?

(f) Step 6: Answer the Question $m\angle x =$ _____ ; $m\angle y =$ _____ ; $m\angle z =$ _____
Note: $m\angle x$ means the measure of angle x.

4. The perimeter *P* of a rectangle is given by the formula $P = 2l + 2w$ where *l* is the length and *w* is the width. Solve the equation for *w* and then use this equation to find the width of a rectangle whose length is 13.5 cm and whose perimeter is 40 cm.

(a) Solve the formula for *w*.

(b) Use part (a) to find the width, *w*, if   4a. _____
the length of the rectangle is 13.5 cm and
the perimeter of the rectangle is 40 cm.   4b. _____

(c) Notice that when solving an applied problem for an unknown in which some of the values of the variables are given, such as in Problem 4, it is possible to solve the equation for the unknown, *w*. Using the new equation, found in 4(a), substitute the given values to answer the question.

It is also possible to substitute the known values into the original equation, $P = 2l + 2w$, and then solve for *w*. Now solve Problem 4 using this approach. Which method do you prefer?

_____

Sullivan/Struve, *Intermediate Algebra*, 3e
Copyright © 2014 Pearson Education, Inc.

Do the Math Exercises 1.3
Using Formulas to Solve Problems

In Problems 1 – 6, solve the formula for the indicated variable.

1. Direct Variation: $y = kx$ for k **2. Algebra:** $y = mx + b$ for m

1. _____

2. _____

3. Conversion: $F = \dfrac{9}{5}C + 32$ for C **4. Geometry:** $A = \dfrac{1}{2}bh$ for h

3. _____

4. _____

5. Sequences: $S - rS = a - ar^n$ for S **6. Trapezoid:** $A = \dfrac{1}{2}h(B + b)$ for b

5. _____

6. _____

In Problems 7 – 10, solve for y.

7. $4x + 2y = 20$ **8.** $\dfrac{2}{3}x - \dfrac{5}{2}y = 5$

7. _____

8. _____

9. $4.8x - 1.2y = 6$ **10.** $\dfrac{2}{5}x + \dfrac{1}{3}y = 8$

9. _____

10. _____

11. Maximum Heart Rate The model $M = -0.85A + 217$ was developed by Miller to determine the maximum heart rate M of an individual who is age A. (SOURCE: Miller et al; "Predicting max HR"; *Medicine and Science in Sports and Exercise*, 25(9), 1077 – 1081)

    (a) Solve the model for A.   **(b)** What is the age of an individual   11a. _____
                                                   whose maximum heart rate is 160?

11b. _____

12. Supplementary Angles Two angles are **supplementary** if the sum of the measures of the angles is 180°. If one angle is twice the measure of its supplement, find the measures of the two angles.

12. _____

13. Complementary Angles Two angles are **complementary** if the sum of the measures of the angles is 90°. If one angle is 30° less than twice its complement, find the measures of the two angles.

13. _____

14. Angles in a Triangle The sum of the measures of the interior angles in a triangle is 180°. The measure of the second angle is 3 times the measure of the first angle. The measure of the third angle is 20° more than the measure of the first angle. Find the measure of each of the interior angles.

14. _____

15. Critical Thinking Suppose that you wish to install a window with the dimensions given in the figure.

8 ft

3 ft

(a) What is the area of the opening of the window?

15a. _____

(b) What is the perimeter of the window?

15b. _____

(c) If glass costs $8.25 per square foot, what is the cost of the glass for the window?

15c. _____

Sullivan/Struve, *Intermediate Algebra*, 3e
Copyright © 2014 Pearson Education, Inc.

Five-Minute Warm-Up 1.4
Linear Inequalities in One Variable

In Problems 1 – 6, replace the question mark by <, >, or = to make the statement true.

1. -0.5 ? 0

1. _____

2. -6.5 ? -7

2. _____

3. $\dfrac{11}{12}$? $\dfrac{13}{16}$

3. _____

4. 2.625 ? $2\dfrac{5}{8}$

4. _____

5. $-1\dfrac{3}{8}$? $-2\dfrac{1}{4}$

5. _____

6. 0.0033 ? 0.0034

6. _____

7. One inequality symbol that is not used in this section is $\neq$ as the symbols used in this section represent strict and nonstrict inequalities.

7a. _____

(a) List the two symbols which represent strict inequalities.

(b) List the two symbols that represent nonstrict (or weak) inequalities.

7b. _____

(c) Use inequality notation to express a variable, x, is between $\dfrac{1}{2}$ and $\dfrac{3}{8}$.

7c. _____

Guided Practice 1.4
Linear Inequalities in One Variable

Objective 1: Represent Inequalities Using the Real Number Line and Interval Notation

1. Which of the following, if any, is appropriate use of inequality notation?

      (a) $5 < x < -1$ **(b)** $-1 \leq x \geq 1$ 1. _____

Objective 2: Understand the Properties of Inequalities

2. *True or False* The Addition Property of Inequality states that the direction of the inequality does not change regardless of the quantity that is added to each side of the inequality.

2. _____

3. *True or False* When multiplying both sides of an inequality by a negative number, we reverse the direction of the inequality symbol.

3. _____

4. *True or False* When multiplying both sides of an inequality by a positive number, we reverse the direction of the inequality symbol.

4. _____

Objective 3: Solve Linear Inequalities *(See textbook Examples 5 and 6)*

5. Solve the inequality $5x + 2 > -13$. Graph the solution set.

Step 1: We isolate the term containing the variable.

                                             $5x + 2 > -13$

Subtract 2 from both sides: **(a)** $5x + 2 \underline{\quad} > -13 \underline{\quad}$

Simplify: **(b)** _____

Step 2: Now we want to get a coefficient of 1 on the variable.

Divide both sides by 5:   **(c)** _____

Simplify: **(d)** _____

Write the answer in set-builder notation: **(e)** _____

Write the answer using interval notation: **(f)** _____

Graph the solution set: **(g)**

6. Solve the inequality $3x + 7 \geq 7x - 1$. Graph the solution set.

$$3x + 7 \geq 7x - 1$$

Subtract 7 from both sides: **(a)** $3x + 7$ _____ $\geq 7x - 1$ _____

Simplify: **(b)** _____

Subtract $7x$ from both sides: **(c)** $3x$ _____ $\geq 7x - 8$ _____

Simplify: **(d)** _____

Divide both sides by -4. Don't forget to change the direction of the inequality symbol. Simplify: **(e)** _____

Write the solution using set-builder notation: **(f)** _____

Write the solution using interval notation: **(g)** _____

Graph the solution set: **(h)**

-4   -2   0   2   4

Objective 4: Solve Problems Involving Linear Inequalities *(See textbook Example 9)*

7. Write the appropriate inequality symbol for each phrase.

Phrase	Inequality Symbol	Phrase	Inequality Symbol
(a) At least	**(a)** _____	**(b)** No less than	**(b)** _____
(c) More than	**(c)** _____	**(d)** Greater than	**(d)** _____
(e) No more than	**(e)** _____	**(f)** At most	**(f)** _____
(g) Fewer than	**(g)** _____	**(h)** Less than	**(h)** _____

8. Commission Nghiep sells digital cameras. His annual salary is $25,000 plus a commission of 20% on all of his sales. What is the value of the digital cameras Nghiep needs to sell so that his annual salary will be at least $36,000?

(a) Step 1: Identify What key word(s) indicates that this requires an inequality? _____

Step 2: Name Let v represent the value of the digital cameras sold.

(b) Step 3: Translate Write an inequality that represents the minimum annual salary. _____

(c) Step 4: Solve

_____

(d) Step 5: Check When you substitute your values into the inequality, is the statement true? _____

(e) Step 6: Answer the Question _____

Sullivan/Struve, *Intermediate Algebra*, 3e
Copyright © 2014 Pearson Education, Inc.

Name:

Instructor:

Date:

Section:

Do the Math Exercises 1.4
Linear Inequalities in One Variable

In Problems 1 – 4, solve each linear inequality. Express your answer in set-builder notation. Graph the solution set.

1. $x + 6 < 9$

2. $5x - 4 \leq 16$

1. _____

2. _____

3. $-7x < 21$

4. $8x + 3 \geq 5x - 9$

3. _____

4. _____

In Problems 5 – 12, solve each linear inequality. Express your answer in interval notation.

5. $3x + 4 \geq 5x - 8$

6. $3(x - 2) + 5 > 4(x + 1) + x$

5. _____

6. _____

7. $2.3x - 1.2 > 1.8x + 0.4$

8. $\dfrac{2x - 3}{3} > \dfrac{4}{3}$

7. _____

8. _____

9. $\dfrac{2}{3} - \dfrac{5}{6}x > 2$

10. $\dfrac{5}{6}(3x - 2) - \dfrac{2}{3}(4x - 1) < -\dfrac{2}{9}(2x + 5)$

9. _____

10. _____

11. $\dfrac{2}{5}x + \dfrac{3}{10} < \dfrac{1}{2}$

12. $\dfrac{x}{12} \geq \dfrac{x}{2} - \dfrac{2x+1}{4}$

11. _____

12. _____

13. Find the set of all x such that the difference between 3 times x and 2 is less than 7.

13. _____

14. Find the set of all y such that the sum of twice y and 3 is greater than 13.

14. _____

15. Computing Grades In order to earn an A in Mrs. Padilla's Intermediate Algebra course, Mark must obtain an average score of at least 90. On his first four exams Mark scored 94, 83, 88, and 92. The final exam counts as two test scores. What score does Mark need on the final to earn an A in Mrs. Padilla's class?

15. _____

16. Moving Trucks A 15 foot moving truck from Budget costs $39.95 per day plus $0.65 per mile. If your budget only allows for you to spend $125.75, what is the maximum number of miles you can drive? (SOURCE: *Budget*)

16. _____

17. Explain why we never mix inequalities as in $4 < x > 7$.

Five-Minute Warm-Up 1.5
Rectangular Coordinates and Graphs of Equations

1. Plot the following points on the real number line: $\dfrac{3}{2}, -2, 0, 3$.

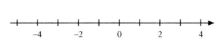

2. Determine which of the following are solutions to the equation $4x - 5(x + 1) = 6$.

   (a) $x = -1$ **(b)** $x = -11$ **(c)** $x = 1$ **(d)** $x = 11$ 2. _____

3. Evaluate the expression $-x^2 - 3x + 4$ for the given values of the variable.

   (a) $x = 3$ **(b)** $x = -2$ **(c)** $x = 0$

   3a. _____

   3b. _____

   3c. _____

4. Solve the equation $4x - 3y = -12$ for y.   4. _____

5. Evaluate each of the following absolute values.   5a. _____

   (a) $\left|-12\right|$ **(b)** $\left|0\right|$ **(c)** $\left|125\right|$ 5b. _____

   5c. _____

Guided Practice 1.5
Rectangular Coordinates and Graphs of Equations

Objective 1: Plot Points in the Rectangular Coordinate System

1. Label each quadrant and axis in the rectangular or Cartesian coordinate system.

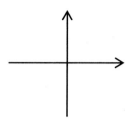

2. What name do we give the ordered pair $(0, 0)$? _____

In Problems 3 and 4, circle one answer for each underlined choice.

3. To plot the ordered pair $(-3, 2)$, you would move 3 units <u>up, down, left, or right</u> from the origin?

4. To plot the ordered pair $(-6, -2)$, you would move <u>6 or 2</u> units down from the origin.

Objective 3: Graph an Equation Using the Point-Plotting Method   *(See textbook Example 3)*

5. Graph the equation $y = 3x - 1$ by plotting points.

Step 1: We want to find all the points (x, y) that satisfy the equation. To determine these points we choose values of x and use the equation to determine the corresponding values of y.

	x	$y = 3x - 1$	(x, y)
(a)	-2	$3(\underline{}) - 1 = \underline{}$	$(-2, \underline{})$
(b)	-1	$3(\underline{}) - 1 = \underline{}$	$(-1, \underline{})$
(c)	0	$3(\underline{}) - 1 = \underline{}$	$(0, \underline{})$
(d)	1	$3(\underline{}) - 1 = \underline{}$	$(1, \underline{})$
(e)	2	$3(\underline{}) - 1 = \underline{}$	$(2, \underline{})$

Step 2: Draw the axes in the Cartesian plane and plot the points listed in the third column. Now connect the points to obtain the graph of the equation (a line).

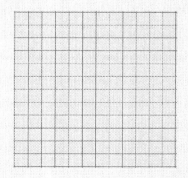

Objective 4: Identify the Intercepts from the Graph of an Equation *(See textbook Example 6)*

6. The point(s), if any, where a graph crosses or touches the *x*-axis is called the *x*-intercept.   6._____
Suppose a graph crosses the *x*-axis at a point *a* and touches the *x*-axis at a point *b*. Write the
x-intercept(s) as an ordered pair.   _____

7. The point(s), if any, where a graph crosses or touches the *y*-axis is called the *y*-intercept.   7._____
Suppose a graph crosses the *y*-axis at a point *c*. Write the *y*-intercept(s) as an ordered pair.

Objective 5: Interpret Graphs

8. $(1, 5)$ is a point on the graph of $3x - y = -2$. If the *x*-axis represents the number of picnic tables
manufactured and sold and the *y*-axis represents the profit (in tens of dollars) from the sale of those tables,
describe the meaning of the ordered pair $(1, 5)$

_____

9. $(2, 8)$ is a point on the graph of $y = -x^2 + 3x + 6$. If the *x*-axis represents the time (in seconds) after a
ball leaves the hand of a thrower and the *y*-axis represents the height (in feet) above the ground, describe the
meaning of the ordered pair $(2, 8)$.

_____

Sullivan/Struve, *Intermediate Algebra*, 3e
Copyright © 2014 Pearson Education, Inc.

Do the Math Exercises 1.5
Rectangular Coordinates and Graphs of Equations

1. Determine the coordinates of each of the points plotted. Tell in which quadrant or on what coordinate axis each point lies.

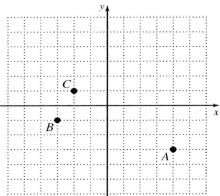

1.

A _____

B _____

C _____

2. Use the graph above to plot each of the following points. Tell in which quadrant or on which coordinate axis each point lies.

$D(0, 5)$ $E(2, 3)$ $F(1, 0)$

2.

D _____

E _____

F _____

3. Determine which of the following are points on the graph of the equation $-4x + 3y = 18$.

(a) $(1, 7)$ **(b)** $(0, 6)$ **(c)** $(-3, 10)$ **(d)** $\left(\dfrac{3}{2}, 4\right)$ 3. _____

4. Determine which of the following are points on the graph of the equation $x^2 + y^2 = 1$.

(a) $(0, 1)$ **(b)** $(1, 1)$ **(c)** $\left(\dfrac{1}{2}, \dfrac{1}{2}\right)$ **(d)** $\left(\dfrac{\sqrt{3}}{2}, \dfrac{1}{2}\right)$ 4. _____

5. The graph of the equation is given. List the intercepts of the graph.

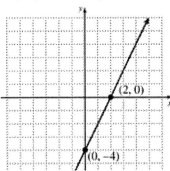

5. *x*-intercept:

_____

y-intercept:

_____

In Problems 6 – 9, graph each equation by plotting points.

6. $y = x - 2$

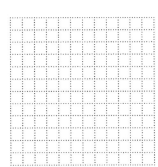

7. $y = x^2 - 2$

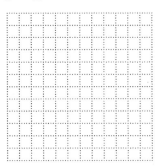

8. $y = |x| + 1$

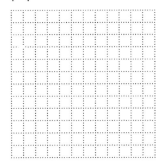

9. $3x + y = -2$

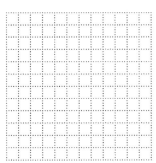

10. If $(a, -2)$ is a point on the graph of $y - 3x = 5$, what is a?

10. _____

11. If $(-2, b)$ is a point on the graph of $y = -2x^2 + 3x + 1$, what is b?

11. _____

Sullivan/Struve, *Intermediate Algebra,* 3e
Copyright © 2014 Pearson Education, Inc.

Five-Minute Warm-Up 1.6
Linear Equations in Two Variables

In Problems 1 and 2, solve each linear equation.

1. $-6x + 36 = 0$ **2.** $5y + 2 = 0$

1. _____

2. _____

In Problems 3 and 4, solve each equation for y.

3. $5x - 2y = -20$ **4.** $\dfrac{2}{3}x - \dfrac{1}{2}y = 1$

3. _____

4. _____

In Problem 5 and 6, evaluate each expression.

5. $\dfrac{-3 - (-7)}{6 - 8}$ **6.** $\dfrac{-1 - 7}{-6 - (-2)}$

5. _____

6. _____

In Problems 7 and 8, use the Distributive Property to simplify each expression.

7. $-3(x - 5)$ **8.** $\dfrac{5}{4}(x - 8)$

7. _____

8. _____

Guided Practice 1.6
Linear Equations in Two Variables

Objective 1: Graph Linear Equations Using Point Plotting

1. When a linear equation in two variables is written in the form $Ax + By = C$ where A, B, and C are real

numbers and both A and B cannot be 0, we say the linear equation is written in _____ form.

Objective 2: Graph Linear Equations Using Intercepts

2. To find the x-intercept(s), if any, of the graph of an equation, let _____ in the equation and solve for x.

3. To find the y-intercept(s), if any, of the graph of an equation, let _____ in the equation and solve for y.

Objective 4: Find the Slope of a Line Given Two Points

4. The ratio of the rise (vertical change) to the run (horizontal change) is called the _____ of the line.

5. The slope of a line which passes through the ordered pairs (x_1, y_1) and (x_2, y_2) is given by the formula:

$$m = \underline{\hspace{3cm}}$$

6. *Properties of Slope* Describe the graph of the line with the following slope:

    (a) Positive _____

    (b) Negative _____

    (c) Zero _____

    (d) Undefined _____

Objective 7: Use the Point-Slope Form of a Line *(See textbook Example 10)*

7. What equation do we use to write the point-slope form of a line with slope m and containing (x_1, y_1)?

        _____

8. Find the equation of a line whose slope is -2 and that contains the point $(4, -1)$.

(a) Identify values: $m = $ _____ ; $x_1 = $ _____ ; $y_1 = $ _____

(b) Point-slope form of a line: _____

(c) Substitute the values into (b):   _____

(d) Simplify: _____

Objective 9: Find the Equation of a Line Given Two Points *(See textbook Example 12)*

9. Find the equation of the line through the points $(2, -2)$ and $(-2, 6)$. If possible, write the equation in slope-intercept form. Graph the line.

Step 1: Find the slope of the line containing the points.	Formula for the slope of a line through two points:	**(a)** _____
	Identify values:	**(b)** $x_1 =$ _____; $y_1 =$ _____
		$x_2 =$ _____; $y_2 =$ _____
	Substitute values into (a) and simplify:	**(c)** _____
Step 2: Use the point-slope form of a line to find the equation.	Point-slope form of a line:	**(d)** _____
	Identify values:	**(e)** $m =$ _____; $x_1 =$ _____; $y_1 =$ _____
	Substitute values into (d):	**(f)** _____
Step 3: Solve the equation for *y*.	Distribute the -2:	**(g)** _____
	Subtract _____ from both sides:	**(h)** _____
	Simplify. Is the equation in slope-intercept form?	**(i)** _____
	Try using (x_2, y_2) in (e). Does this change the equation of the line found in (i)?	**(j)** _____
	Identify the slope:	**(k)** _____
	Identify the *y*-intercept:	**(l)** _____
	Graph the line:	**(m)**

Sullivan/Struve, *Intermediate Algebra*, 3e
Copyright © 2014 Pearson Education, Inc.

Do the Math Exercises 1.6
Linear Equations in Two Variables

In Problems 1 – 4, graph each linear equation in two variables.

1. $x + y = -3$

2. $\dfrac{1}{4}x + \dfrac{1}{5}y = 1$

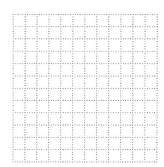

3. $x = 2$

4. $y = -4$

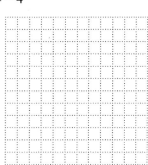

In Problems 5 and 6, determine the slope of the line containing the two points.

5. $(3, -1)$ and $(-2, 11)$

6. $(4, 1)$ and $(4, -3)$

5. _____

6. _____

7. Graph the line containing the point $P(-1, 4)$ and having slope $m = 2$.

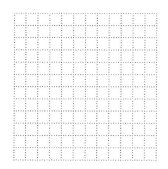

In Problems 8 – 10, find an equation of the line with the given slope and containing the given point. Express your answer in slope-intercept form.

8. $m = -1;\ (0, 0)$

9. $m = 4;\ (2, -1)$

8. _____

9. _____

10. $m = -\dfrac{4}{3}, (1, -3)$

10. _____

11. Find the equation of the line containing $(-5, 1)$ and $(1, -1)$. Express your answer in slope-intercept form.

11. _____

In Problems 12 – 15, find the slope and y-intercept of each line.

12. $y = 3x + 2$ **13.** $2x + y = 3$

12. _____

13. _____

14. $x = 3$ **15.** $y = -4$

14. _____

15. _____

16. Find a linear function g such that $g(1) = 5$ and $g(5) = 17$. What is $g(-3)$?

16. function:

_____

$g(-3) = $ _____

17. Find a linear function F such that $F(2) = 5$ and $F(-3) = 9$. What is $F\left(-\dfrac{3}{2}\right)$?

17. function:

_____

$F\left(-\dfrac{3}{2}\right) = $ _____

18. What type of line has one x-intercept but no y-intercept?

18. _____

19. What type of line has one y-intercept but no x-intercept?

19. _____

Sullivan/Struve, *Intermediate Algebra*, 3e
Copyright © 2014 Pearson Education, Inc.

Five-Minute Warm-Up 1.7
Parallel and Perpendicular Lines

1. Determine the reciprocal of 1.

2. Determine the reciprocal of -5.

1. _____

2. _____

3. Determine the reciprocal of $-\dfrac{3}{2}$.

4. Determine the reciprocal of $\dfrac{1}{3}$.

3. _____

4. _____

5. Solve for y: $-7x + 3y = 9$

6. Find the slope of the line: $4x + y = 3$

5. _____

6. _____

7. Find the slope of the line: $x = -7$

8. Find the slope of the line: $y = 3$

7. _____

8. _____

Guided Practice 1.7
Parallel and Perpendicular Lines

Objective 1: Define Parallel Lines

1. In your own words, write a definition of parallel lines.

2. Two lines are parallel if and only if they have the same _____ and different _____.

3. If $L_1 : y = m_1 x + b_1$ and $L_2 : y = m_2 x + b_2$ and L_1 is parallel to L_2, then m_1 _____ m_2 and b_1 _____ b_2.

Objective 2: Find Equations of Parallel Lines *(See textbook Example 2)*

4. Find an equation for the line that is parallel to $3x - 4y = 2$ and contains the point $(4, -1)$. Graph the lines in the Cartesian plane.

Step 1: Find the slope of the given line by putting the equation in slope-intercept form.

(a)_____

Step 2: Use the point-slope form of a line with the given point and the slope found in Step 1 to find the equation of the parallel line.

Slope of the line found in Step 1:   **(b)**_____

Slope of the parallel line: **(c)**_____

Identify values: **(d)** $m =$ _____; $x_1 =$ _____; $y_1 =$ _____

Point-slope form of a line: **(e)**_____

Substitute values into (e):   **(f)**_____

Step 3: Put the equation in slope-intercept form by solving for *y*.

Solve for *y*:   **(g)**_____

Graph both lines in the Cartesian plane: **(h)**

Objective 3: Define Perpendicular Lines

5. In your own words, write a definition of perpendicular lines.

6. Two lines are perpendicular if and only if the product of their slopes is _____.

7. If $L_1 : y = m_1 x + b_1$ and $L_2 : y = m_2 x + b_2$ and L_1 is perpendicular to L_2, then $m_1 \cdot m_2 =$ ____ or $m_1 =$ ____.

Objective 4: Find Equations of Perpendicular Lines *(See textbook Example 5)*

8. Find an equation of the line that is perpendicular to the line $3x - y = 6$ and contains the point $(-9, 1)$. Write the equation in slope-intercept form.

Step 1: Find the slope of the given line by putting the equation in slope-intercept form.

(a)_____

Step 2: Find the slope of the perpendicular line.

Slope of the line found in Step 1: **(b)**_____

Slope of the perpendicular line: **(c)**_____

Identify values: **(d)** $m =$ ____ ; $x_1 =$ ____ ; $y_1 =$ ____

Step 3: Use the point-slope form of a line with the given point and the slope found in Step 2 to find the equation of the perpendicular line.

Point-slope form of a line: **(e)**_____

Substitute values into (e): **(f)**_____

Step 4: Put the equation in slope-intercept form by solving for *y*.

Solve for *y*: **(g)**_____

Graph both lines in the Cartesian plane: **(h)**

Sullivan/Struve, *Intermediate Algebra*, 3e
Copyright © 2014 Pearson Education, Inc.

Do the Math Exercises 1.7
Parallel and Perpendicular Lines

1. *Complete the following chart:*

	Slope of the given line	Slope of a line parallel to the given line	Slope of a line perpendicular to the given line
a.	-10		
b.		2	
c.			$-\dfrac{2}{3}$
d.	0		
e.			0

In Problems 2 and 3, without graphing determine whether the given linear equations are parallel, perpendicular, or neither.

2. $L_1 : -3x - y = 3$
 $L_2 :\ \ x - 3y = 12$

3. $L_1 : 10x - 3y = 5$
 $L_2 : 5x + 6y = 3$

2. _____

3. _____

4. Determine whether the lines containing the following pairs of points are parallel, perpendicular, or neither.

$\qquad$ L_1: (1, −3); (5, −4) and L_2: (−1, −3); (−2, −7)

4. _____

In Problems 5 and 6, find an equation of the line with the given properties. Express your answer in slope-intercept form.

5. Parallel to $y = -3x + 1$ through the point (2, 5)

6. Perpendicular to $y = 4x + 3$ through the point (4, 1)

5. _____

6. _____

In Problems 7 – 12, find an equation of the line with the given properties. Express your answer in slope-intercept form.

7. Parallel to $x = -2$ through the point $(2, 5)$

8. Perpendicular to $-2x + 5y - 3 = 0$ through the point $(2, -3)$

7. _____

8. _____

9. Parallel to $2x + y = 5$ through the point $(-4, 3)$

10. Perpendicular to $y = 8$ through the point $(2, -4)$

9. _____

10. _____

11. Perpendicular to $3x + y = 1$ through the point $(3, -1)$

12. Parallel to the line $x + 4y = 2$ and through the point $(-7, 2)$

11. _____

12. _____

13. Find B so that $-6x + By = 3$ is perpendicular to $2x - 3y = 8$.

13. _____

14. Geometry In geometry, we learn that a parallelogram is a quadrilateral in which both pairs of opposite sides are parallel. Determine whether the points $A = (-2, -1)$, $B = (4, 1)$, $C = (5, 5)$, and $D = (-1, 3)$ are the vertices of a parallelogram. What is the slope of each of the sides of the quadrilateral?

14. slope of:

$\overline{AB}$ _____

$\overline{BC}$ _____

$\overline{CD}$ _____

$\overline{AD}$ _____

Is ABCD a parallelogram?_____

Five-Minute Warm-Up 1.8
Linear Inequalities in Two Variables

In Problems 1 and 2, determine whether the given value satisfies the inequality. (Yes or No)

1. $-4x + 1 \geq -3;\ x = 1$

2. $-y - 15 < 3(2y - 1):\ y = -2$

1. _____

2. _____

In Problems 3 – 6, solve the inequality in one variable.

3. $2x - 2 \geq 3 + x$

4. $8 - 4(2 - x) \leq -2x$

3. _____

4. _____

5. $n + 3 > \dfrac{5}{2}(n - 6)$

6. $\dfrac{a}{6} < 2 + \dfrac{a}{3}$

5. _____

6. _____

Guided Practice 1.8
Linear Inequalities in Two Variables

Objective 1: Determine Whether an Ordered Pair Is a Solution to a Linear Inequality

1. *True or False* To determine if an ordered pair is a solution to the linear inequality, substitute the values for the variables into the inequality. If a true statement results, then the ordered pair is a solution to the inequality.

1. _____

Objective 2: Graph Linear Inequalities *(See textbook Example 2)*

2. To graph any linear inequality in two variables, you must first graph the corresponding equation.

    (a) If the inequality is strict ($<$ or $>$), you should use a _____ line.

    (b) If the inequality is nonstrict $(\leq$ or $\geq)$, you should use a _____ line.

3. The graph of a line separates the *xy*-plane into two _____.

4. If a test point satisfies the inequality, then every ordered pair that lies in that half plane also satisfies the inequality. To represent this solution set, we _____ the half plane containing the test point.

5. Graph the linear inequality $3x - 4y > 12$.

Step 1: We replace the inequality symbol with an equal sign and graph the corresponding line.	Write the equation:	**(a)** _____
	Identify the *x*-intercept:	**(b)** _____
	Identify the *y*-intercept:	**(c)** _____
	Graph the line. Is the line connecting these points solid or dashed?	**(d)** _____
Step 2: We select any test point that is not on the line and determine whether the test point satisfies the inequality. When the line does not contain the origin, it is usually easiest to choose the origin, (0, 0), as the test point.	Select a test point:	**(e)** _____ $3x - 4y > 12$
	Substitute the values for the variables into the inequality. Is the statement true or false?	**(f)** _____
	Which half plane should be shaded; the half plane containing the test point or the opposite half plane?	**(g)** _____

Exercise 5 continued…

Graph the solution set by plotting the intercepts, graphing the line, and shading the appropriate half plane.   **(h)** $3x - 4y > 12$

Objective 3: Solve Problems Involving Linear Inequalities *(See textbook Example 4)*

6. Salesperson Juanita sells two different computer models. For each Model A computer sold she makes $45 and for each Model B computer sold she makes $65. Juanita set a monthly goal of earning at least $4000.

(a) Write a linear inequality that describes Juanita's options for making her sales goal.

   Step 1: Identify We want to determine how many of each model should be sold so that she will earn at least $4000. This requires an inequality in two variables.

   Step 2: Name the Unknowns Let x represent the number of Model A sold and let y represent the number of Model B sold.

   Step 3: Translate If Model A sells for $45 each and Model B sells for $65 each, write an inequality the represents her total income greater than or equal to $4000.

   (a) _____

(b) Will Juanita makes her sales goal if she sells 50 Model A and 28 Model B computers? _____

(c) Will Juanita makes her sales goal if she sells 41 Model A and 33 Model B computers? _____

Sullivan/Struve, *Intermediate Algebra*, 3e
Copyright © 2014 Pearson Education, Inc.

Do the Math Exercises 1.8
Linear Inequalities in Two Variables

In Problems 1 and 2, determine which, if any, of the following points are solutions to the linear inequality.

1. $2x + y > -3$

    (a) $(2, -1)$ **(b)** $(1, -3)$ **(c)** $(-5, 4)$

2. $2x - 5y \le 2$

    (a) $(1, 2)$ **(b)** $(3, 0)$ **(c)** $(-3, -2)$

1. _____

2. _____

In Problems 3 – 8, graph each inequality.

3. $y < -2$

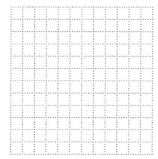

4. $x \ge -3$

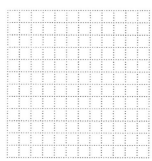

5. $y \ge -\dfrac{4}{3}x + 5$

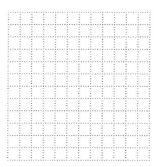

6. $-4x + y \ge -5$

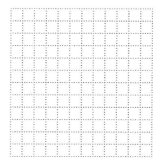

7. $\dfrac{x}{3} - \dfrac{y}{4} > 1$

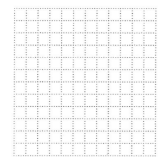

8. $-4x + 6y > 24$

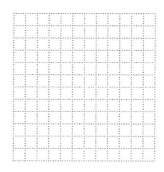

In Problems 9 – 13, translate each statement into a linear inequality.

9. One number, x, is at most 12 more than a second number, y.

9. _____

10. The sum of two numbers, x and y, is at least -3.

10. _____

11. Budget Constraints Johnny can spend no more than $3.00 that he got from his grandparents. He goes to the candy store and wants to buy gummy bears that cost $0.10 each and suckers that cost $0.25 each.

(a) Write a linear inequality that describes Johnny's options for buying candy. Let g represent the number of gummy bears and s represent the number of suckers.

11a. _____

(b) Can Johnny buy 18 gummy bears and 5 suckers?

11b. _____

(c) Can Johnny buy 19 gummy bears and 4 suckers?

11c. _____

12. Fund Raising For a fund-raiser, the Math club agrees to sell candy bars and candles. The club's profit will be 50¢ for each candy bar it sells and $2.00 for each candle it sells. The club needs to earn at least $1000 in order to pay for an upcoming field trip.

(a) Write a linear inequality that describes the combination of candy bars and candles that must be sold. Let x represent the number of candy bars sold and y represent the number of candles sold.

12a. _____

(b) Will selling 500 candy bars and 350 candles earn enough for the trip?

12b. _____

(c) Will selling 600 candy bars and 400 candles earn enough for the trip?

12c. _____

13. Determine the linear inequality whose graph is:

13. _____

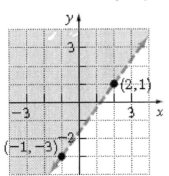

Sullivan/Struve, *Intermediate Algebra*, 3e
Copyright © 2014 Pearson Education, Inc.

Name:

Instructor:

Date:

Section:

Five-Minute Warm-Up 2.1
Relations

1. Write each inequality using interval notation.

  (a) $-1 < x \le 0$ **(b)** $4 \le x < 7$

1a. _____

1b. _____

2. Write each inequality using interval notation.

  (a) $x \le -4$ **(b)** $x > -2$

2a. _____

2b. _____

3. Plot the ordered pairs in the rectangular coordinate system.

$$A(0, -3);\ B(-2, -4);\ C(1, -1);\ D(2, 0);\ E(-4, 3)$$

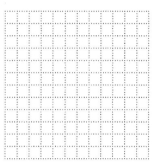

4. Graph the equation $5x + 3y = -15$.

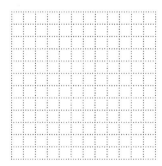

5. Graph the equation $y = -x^2 + 2$ by plotting points.

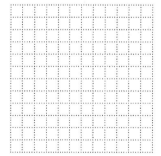

Guided Practice 2.1
Relations

Objective 1: Understand Relations

1. In your own words, write a definition for a *relation*.

Objective 2: Find the Domain and the Range of a Relation *(See textbook Example 4)*

2. The *domain* is set of all _____ and is set of _____ -coordinates for the relation which is defined by the set of ordered pairs (x, y).

3. The *range* is set of all _____ and is set of _____ -coordinates for the relation which is defined by the set of ordered pairs (x, y).

4. Name four different ways to represent a relation.

_____ _____ _____ _____

*In Problems 5 and 6, the figure shows the graph of a relation. Determine **(a)** the domain and **(b)** the range of the relation.*

5.   **6.**

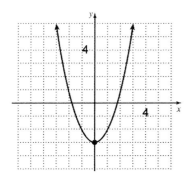   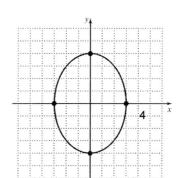

5a. _____

5b. _____

6a. _____

6b. _____

Objective 3: Graph a Relation Defined by an Equation *(See textbook Example 5)*

7. Graph the relation $x = y^2 - 3$. Use the graph to determine **(a)** the domain and **(b)** the range of the relation.

$x = y^2 - 3$	y	(x, y)
	-2	
	-1	
	0	
	1	
	2	

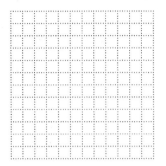

7a. _____

7b. _____

Do the Math Exercises 2.1
Relations

Write the relation as a set of ordered pairs. Then identify the domain and range.

1.

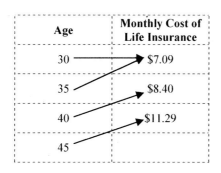

Age	Monthly Cost of Life Insurance
30	$7.09
35	$8.40
40	$11.29
45	

1. _____

domain _____

range _____

In Problems 2 – 4, identify the domain and the range of each relation.

2. $\{(-2, 6), (-1, 3), (0, 0), (1, -3), (2, 6)\}$

    domain _____

    range _____

3. $\{(-2, -8), (-1, -1), (0, 0), (1, 1), (2, 8)\}$

    domain _____

    range _____

4. $\{(-3, 0), (0, 3), (3, 0), (0, -3)\}$

    domain _____

    range _____

In Problems 5 – 8, identify the domain and range of the relation from the graph.

5.

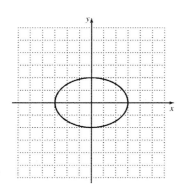

    domain _____

    range _____

6.

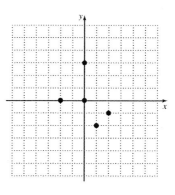

    domain _____

    range _____

7.

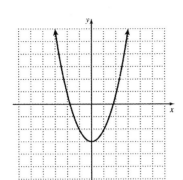

domain _____

range _____

8.

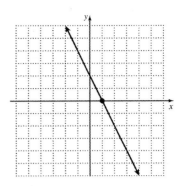

domain _____

range _____

In Problems 9 – 14, identify the domain and range of each relation. You may want to draw a graph by point-plotting or by using your graphing calculator.

9. $y = x - 2$

domain _____

range _____

10. $y = x^2 - 2$

domain _____

range _____

11. $y = -2x^2 + 8$

domain _____

range _____

12. $y = |x| - 2$

domain _____

range _____

13. $x = y^2 + 2$

domain _____

range _____

14. $y = x^3 - 4$

domain _____

range _____

15. Bob Villa wishes to put a new window in his home. He wants the perimeter of the window to be 100 feet. The graph shows the relation between the width, *x*, of the opening and the area of the opening.

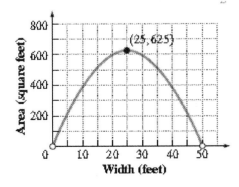

(a) Determine the domain and the range of the relation.

15a._____

_____

(b) Explain why the domain obtained in part (a) is reasonable.

_____

_____

Five-Minute Warm-Up 2.2
An Introduction to Functions

In Problems 1 and 2, evaluate the expression for the given value of the variable.

1. $-\dfrac{4}{5}x - 3$ for $x = -25$ **2.** $-2x^2 + 3x - 1$ for $x = -1$

1. _____

2. _____

3. Express the inequality $x > -16$ using interval notation.

3. _____

4. Express the interval $(-\infty, 100]$ using set-builder notation.

4. _____

5. For the following set of ordered pairs, list **(a)** the domain and **(b)** the range.
$$\{(-2, 1), (-4, 3), (-6, 5), (-8, 7)\}$$

5a. domain _____

5b. range _____

In Problems 6 – 8, evaluate each expression.

6. $\dfrac{2^2 - 5}{-5}$ **7.** $-4(-1)^2 + 6(-1) + 5$ **8.** $\left| \dfrac{4 - (-20)}{-2^3} \right|$

6. _____

7. _____

8. _____

9. The volume of a right circular cylinder is given by the formula $V = \pi r^2 (r + 3)$ where r
is the radius of the cylinder and whose height is 3 inches more than the radius. Find the
volume of a cylinder whose radius is 2.5 inches. Round your answer to the nearest tenth
of a cubic inch.

9. _____

Guided Practice 2.2
An Introduction to Functions

Objective 1: Determine Whether a Relation Expressed as a Map or Ordered Pairs Represents a Function

1. Explain what a *function* is. Be sure to include the terms *domain* and *range* in your explanation.

2. Determine whether the relation represents a function. If the relation is a function, then state its domain and range. *(See textbook Example 2)*

    (a) $\{(3, -6), (4, -1), (5, -6)\}$ _____

    (b) $\{(4, 5), (3, -3), (4, -1)\}$ _____

Objective 2: Determine Whether a Relation Expressed as an Equation Represents a Function

3. The symbol $\pm$ is a shorthand device and is read "plus or minus." Write the two equations that are represented by $y = \pm 2x$ and then determine whether the equation $y = \pm 2x$ is a function.

    (a) $y = \pm 2x$ means $y =$ _____ and also $y =$ _____ . **(b)** Is y a function of x? _____

4. Determine whether each equation shows y as a function of x. *(See textbook Examples 3 and 4)*
    (a) $y = 9$ **(b)** $x + y^2 = 2$

                                                    4a. _____

                                                    4b. _____

Objective 3: Determine Whether a Relation Expressed as a Graph Represents a Function

5. We use the Vertical Line Test to determine whether the graph of a relation is a function. In your own words, state the *Vertical Line Test (VLT)*.

6. Which of the following are graphs of functions? *(See textbook Example 5)*   6. _____

    (a)   **(b)**   **(c)**   **(d)**

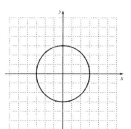
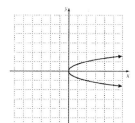
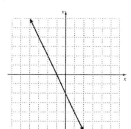
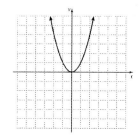

Objective 4: Find the Value of a Function

7. For the function $y = f(x)$, the variable x is called the _____ variable, because it can be

assigned any of the values in the _____.

8. For the function $y = f(x)$, the variable y is called the _____ variable, because it can be

assigned any of the values in the _____.

In Problems 9 – 12, find the value of each function. (See textbook Example 6 and Example 7)

9. $f(x) = 2x^2 - 5x$; $f(-4)$ 10. $g(x) = -3x + 2$; $g(2a)$

9. _____

10. _____

11. $h(t) = 4$; $h(5)$ 12. $F(z) = -3z + 4$; $F(k + 1)$

11. _____

12. _____

Objective 5: Find the Domain of a Function

13. The domain of a function is the set of all inputs for which the function gives an output that is a real
    number or makes sense. When identifying the domain of a function we exclude values of the variable
    that causes division by _____.

14. Find the domain of the function: $f(x) = \dfrac{2x}{2x + 1}$ *(See textbook Example 8)*   14. _____

Objective 6: Work with Applications of Functions

15. The number N of trucks produced at a certain factory in 1 day after t hours of operation is
    given by $N(t) = 80t - 4t^2$, where $0 \le t \le 8$. *(See textbook Example 10)*

    (a) Identify the dependent variable.   15a. _____

    (b) Identify the independent variable.   15b. _____

    (c) Evaluate $N(5)$ and explain the meaning of $N(5)$.   15c. _____

Name: Date:
Instructor: Section:

Do the Math Exercises 2.2
An Introduction to Functions

In Problems 1 – 3, determine whether the relation represents a function (Yes or No).

1.

Grade on exam	Study time (hr)
A	5
B	6
C	4
D	3.5
	1

1. _____

2. $\{(-2, 3), (-2, 1), (-2, -3), (-2, 9)\}$

2. _____

3. $\{(-5, 3), (-2, 1), (5, 1), (7, -3)\}$

3. _____

In Problems 4 – 7, determine whether each equation shows y as a function of x. (Yes or No)

4. $y = -6x + 3$ **5.** $y = \pm 4x$

4. _____

5. _____

6. $y = x^2 + 2$ **7.** $y^2 = x$

6. _____

7. _____

In Problems 8 – 11, determine whether the graph is that of a function (Yes or No).

8.

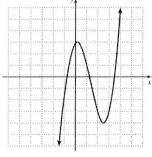

9.

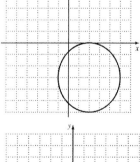

8. _____

9. _____

10.

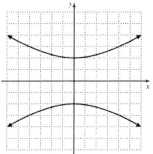

11.

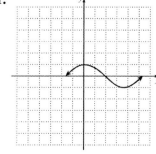

10. _____

11. _____

In Problems 12 and 13, find the following values for each function.

(a) $f(3)$ (b) $f(-2)$ (c) $f(2x)$ (d) $f(x+2)$

12. $f(x) = 3x + 1$ **13.** $f(x) = -2x - 3$

12a. _____

12b. _____

12c. _____

12d. _____

13a. _____

13b. _____

13c. _____

13d. _____

In Problems 14 and 15, find the domain of each function.

14. _____

14. $f(x) = -2x^2 + x + 1$ **15.** $h(q) = \dfrac{3q^2}{q+2}$

15. _____

16. If $f(x) = -2x^2 + 5x + C$ and $f(-2) = -15$, what is the value of C?   16. _____

17. If $f(x) = \dfrac{-x+B}{x-5}$ and $f(3) = -1$, what is the value of B?   17. _____

Sullivan/Struve, *Intermediate Algebra*, 3e
Copyright © 2014 Pearson Education, Inc.

Five-Minute Warm-Up 2.3
Functions and Their Graphs

In Problems 1 and 2, solve each equation.

1. $-4x + 2 = 0$ **2.** $-18 + 3y = 0$ 1. _____

                                                                        2. _____

3. Graph the equation $y = -\dfrac{4}{3}x + 2$.   **4.** Graph the equation $y = x^2 - 4$ by plotting points.

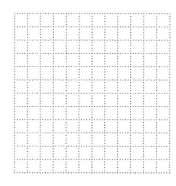   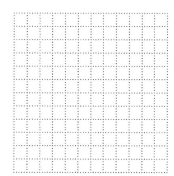

*In Problems 5 and 6, determine the **(a)** domain and **(b)** the range from the graph.*

5.   **6.**   5a. _____

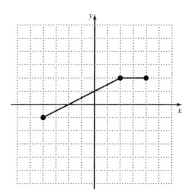   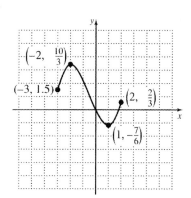   5b. _____

                                                                                     6a. _____

                                                                                     6b. _____

Guided Practice 2.3
Functions and Their Graphs

Objective 1: Graph a Function

1. Complete the table and graph the function $f(x) = |2x - 4|$. *(See textbook Example 1)*

x	$f(x)$	$(x, f(x))$
-2		
-1		
0		
1		
2		
3		
4		

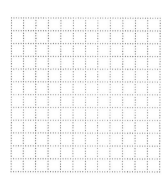

Objective 2: Obtain Information from the Graph of a Function

2. Use the graph of the function to answer parts (a) – (c). *(See textbook Example 2)*

   (a) Determine the domain of the function.   2a. _____

   (b) Determine the range of the function.   2b. _____

   (c) Identify the x-intercept(s).   2c. _____

   (d) Identify the y-intercept(s).   2d. _____

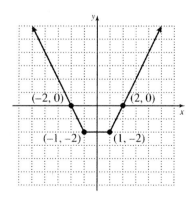

3. Consider the function $f(x) = -\dfrac{5}{2}x + 8$. *(See textbook Example 4)*

    (a) Is the point $(4, -2)$ on the graph of the function?

3a. _____

    (b) If $x = 6$, what is $f(x)$? What point is on the graph of the function?

3b. _____

_____

    (c) If $f(x) = 3$, what is x? What point is on the graph of f?

3c. _____

_____

4. The zeros of a function are also the ____-intercepts of the graph of the function. To find the zeros of a function we set the function equal to _____ and solve for x.

5. Find the zeros of the function f whose graph is shown below. *(See textbook Example 5)*

5. _____

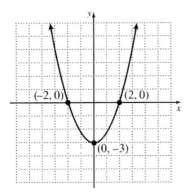

Objective 3: Know Properties and Graphs of Basic Functions

6. List the 6 basic functions described in Table 3 and briefly describe each graph.

    (a) _____

    (b) _____

    (c) _____

    (d) _____

    (e) _____

    (f) _____

Do the Math Exercises 2.3
Functions and Their Graphs

In Problems 1–5, find the domain of each function.

1. $G(x) = -8x + 3$

2. $H(x) = \dfrac{x+5}{2x+1}$

3. $s(t) = 2t^2 - 5t + 1$

4. $H(q) = \dfrac{1}{6q+5}$

5. $f(x) = \dfrac{4x-9}{7}$

1. _____

2. _____

3. _____

4. _____

5. _____

In Problems 6 and 7, graph each function.

6. $F(x) = x^2 + 1$

7. $H(x) = |x+1|$

In Problems 8 and 9, find (a) the domain, (b) the range, (c) the intercepts, if any, and (d) the zeros, if any.

8.

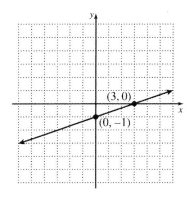

9.

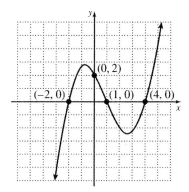

8a. _____

8b. _____

8c. _____

8d. _____

9a. _____

9b. _____

9c. _____

9d. _____

10. *Use the table of values for the function G to answer the following:*

x	G(x)
−5	−3
−4	0
0	5
3	8
7	5

    (a) What is $G(3)$?

10a. _____

    (b) For what number(s) is $G(x) = 5$?

10b. _____

    (c) What is the *x*-intercept of the graph of *G*?

10c. _____

    (d) What is the *y*-intercept of the graph of *G*?

10d. _____

    (e) Are there any zeros of the function *G*? If so, name the zero(s).

10e. _____

11. *Use the function $f(x) = 3x + 5$ to answer the following:*

    (a) Is the point (−2, 1) on the graph of the function?

    (b) If $x = 4$, what is $f(x)$? What point is on the graph of the function?

11a. _____

11b. _____

    (c) If $f(x) = -4$, what is *x*? What point is on the graph of the function?

    (d) Is $x = 0$ a zero of the function?

11c. _____

11d. _____

12. Geometry The volume *V* of a sphere as a function of its radius *r* is given by $V(r) = \frac{4}{3}\pi r^3$.

    (a) What is the domain of this function?

12a. _____

    (b) Find the volume of the sphere whose radius is $4\frac{1}{2}$ cm. Round your answer to one decimal place.

12b. _____

13. A **piecewise-defined function** is a function defined by more than one equation. For example, the absolute value function $f(x) = |x|$ is actually defined by two equations: $f(x) = x$ if $x \geq 0$ and $f(x) = -x$ if $x < 0$. We can combine these equations into one piecewise-defined function written as

$$f(x) = \begin{cases} x & \text{for } x \geq 0 \\ -x & \text{for } x < 0 \end{cases}$$

To evaluate $f(3)$, we recognize that $3 \geq 0$, so we use the rule $f(x) = x$ and obtain $f(3) = 3$. To evaluate $f(-4)$, we recognize that $-4 < 0$, so we use the rule $f(x) = -x$ and obtain $f(-4) = -(-4) = 4$.

Given $f(x) = \begin{cases} x + 3 & \text{for } x < 0 \\ -2x + 1 & \text{for } x \geq 0 \end{cases}$, find each of the following:

    (a) $f(3)$ **(b)** $f(-2)$ **(c)** $f(0)$

13a. _____

13b. _____

13c. _____

Five-Minute Warm-Up 2.4
Linear Functions and Models

In Problems 1 – 2, graph each linear equation.

1. $\dfrac{2}{3}x - \dfrac{1}{2}y = -2$

2. $x = -1$

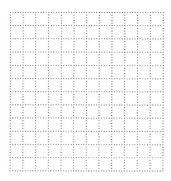

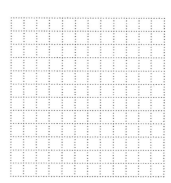

3. Find the equation of the line through $(-1,\ 2)$ and $(2, -7)$.

3. _____

In Problems 4 and 5, solve for the given variable.

4. $6.5 - 1.5(x - 4) + 4x = 10.25$

5. $-\dfrac{7}{8}(2y + 3) \le \dfrac{5}{4}(y - 2)$

4. _____

5. _____

Guided Practice 2.4
Linear Functions and Models

Objective 1: Graph Linear Functions

1. A _____ function is a function of the form $f(x) = mx + b$ where m and b are real numbers. The

graph of a linear function is called a _____.

2. To graph a linear function, we use the same technique used to graph a linear equation written in

slope-intercept form, $y = mx + b$, where m is the _____ and $(0, b)$ is the _____.

3. Graph the linear function $f(x) = \dfrac{3}{2}x - 2$. *(See textbook Example 1)*

    (a) Identify the *y*-intercept.   3a. _____

    (b) Identify the slope.   3b. _____

    (c) Graph the function.

Objective 2: Find the Zero of a Linear Function

4. To find the zero of a linear function $f(x) = mx + b$, we solve the equation _____.

5. Perimeter of a Rectangle In a given rectangle, the length is 3 ft less than twice the width. If x represents the width of the rectangle, the perimeter can be calculated by the function $P(x) = 2x + 2(2x - 3)$. *(See textbook Example 3)*

    (a) What is the implied domain of the function?   5a. _____

    (b) What is the perimeter of a rectangle whose width is 12 ft?   5b. _____

    (c) What is the width of a rectangle whose perimeter is 84 ft?   5c. _____

    (d) For what width of the rectangle will the perimeter exceed 12 feet?   5d. _____

Objective 3: Build Linear Models from Verbal Descriptions

6. The linear cost function is $C(x) = ax + b$, where b represents the _____ costs of operating a

business and _____ represents the costs associated with manufacturing one additional item. *(See textbook Example 4)*

7. Some companies use *straight-line depreciation* to depreciate their assets so that the value of the asset declines by a constant amount each year. To calculate the amount the asset depreciates each year, divide the total cost by the number of years of useful life.

A cab company bought a new car for $22,500 and plans to drive it until there is no scrap value. The life of a car in the cab fleet is 5 years. *(See textbook Example 5)*

    (a) By how much does the car depreciate each year?   7a. _____

    (b) This rate can be expressed as the slope of a linear function. Is this slope positive or negative?   7b. _____

    (c) The *book value* is the value of the asset at a particular time. To find the book value, we take the original value and deduct the amount of depreciation after a given time.

    Write a linear function that expresses the book value V of the car as a function of its age, x.   7c. _____

    (d) What is the implied domain of the linear function?   7d. _____

    (e) What is the book value after 4 years?   7e. _____

    (f) When will the book value of the car be $10,125?   7f. _____

    (g) What is the independent variable?   7g. _____

    (h) What is the dependent variable?   7h. _____

Objective 4: Build Linear Models from Data

8. How many points are needed to determine the equation of a line?   8. _____

9. If the data appears to be linearly related, we select two points on the line of best fit. The linear model can be determined by finding the equation of the line through the two points as described in Section 1.6, Example 12.

    (a) To find the equation of the line, you must first determine the _____.

    (b) Second, use the point-slope form of a line, _____, to find the equation.

Sullivan/Struve, *Intermediate Algebra*, 3e
Copyright © 2014 Pearson Education, Inc.

Name:
Instructor:

Date:
Section:

Do the Math Exercises 2.4
Linear Functions and Models

In Problems 1 – 4, graph each linear function.

1. $f(x) = 2x - 4$

2. $g(x) = -\dfrac{3}{2}x + 1$

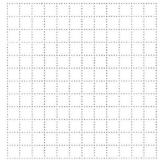 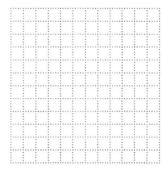

3. $h(x) = \dfrac{1}{4}x + 2$

4. $F(x) = 3$

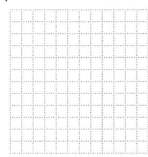 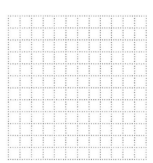

In Problems 5 and 6, find the zeros of the linear function.

5. $f(x) = 3x - 24$

6. $g(x) = -\dfrac{3}{2}x + 6$

5. _____

6. _____

7. Suppose that $f(x) = \dfrac{4}{3}x + 5$ and $g(x) = \dfrac{1}{3}x + 1$.

   (a) Solve $f(x) = g(x)$.

7a. _____

   (b) What is the value of f at the solution?

7b. _____

   (c) What is the value of g at the solution?

7c. _____

   (d) Solve $f(x) > g(x)$.

7d. _____

8. Graph $f(x) = \dfrac{4}{3}x + 5$ and $g(x) = \dfrac{1}{3}x + 1$ on the same Cartesian plane. Label the intersection point.

9. (a) Find a linear function g such that $g(1) = 5$ and $g(5) = 17$. **(b)** What is $g(-3)$?

9a. _____

9b. _____

10. Birth Rate A multiple birth is any birth with 2 or more children born. The birth rate is the number of births per 1,000 women. The birth rate B of multiple births as a function of age a is given by the function $B(a) = 1.73a - 14.56$ for $15 \leq a \leq 44$. [Source: Centers for Disease Control]

(a) What is the independent variable?

10a. _____

(b) What is the dependent variable?

10b. _____

(c) What is the domain of this linear function?

10c. _____

(d) What is the multiple birth rate of women who are 22 years of age according to the model?

10d. _____

(e) What is the age of women whose multiple birth rate is 49.45?

10e. _____

11. A strain of *E. coli* Beu 397-recA441 is placed into a Petri dish at 30° Celsius and allowed to grow. The population is estimated by means of an optical device in which the amount of light that passes through the Petri dish is measured. The data below was collected. Do you think that a linear function could be used to describe the relation between the two variables? Why or Why not?

11. _____

Time, x	Population, y
0	0.09
2.5	0.18
3.5	0.26
4.5	0.35
6	0.50

_____

_____

_____

Do the Math Exercises 2.5
Compound Inequalities

In Problems 1 – 3, use $A = \{4, 5, 6, 7, 8, 9\}$, $B = \{1, 5, 7, 9\}$, *and* $C = \{2, 3, 4, 6\}$ *to find each set.*

1. $A \cup C$ **2.** $A \cap C$ 1. _____

                                                                              2. _____

3. $B \cap C$ 3. _____

In Problems 4 and 5, use $E = \{x \,|\, x \le 2\}$ *and* $F = \{x \,|\, x \ge -2\}$ *to find each of the following.*

4. $E \cap F$ **5.** $E \cup F$ 4. _____

                                                                              5. _____

In Problems 6 – 17, solve each compound inequality. Express your answer in interval notation.

                                                                              6. _____

6. $x \le 5$ and $x > 0$ **7.** $x < 0$ or $x \ge 6$ 7. _____

                                                                              8. _____

8. $7x + 2 \ge 9$ and $4x + 3 \le 7$ **9.** $x + 3 \le 5$ or $x - 2 \ge 3$ 9. _____

                                                                              10. _____

10. $-12 < 7x + 2 \le 6$ **11.** $3x \ge 7x + 8$ or $x < 4x - 9$ 11. _____

                                                                              12. _____

12. $-\dfrac{4}{5}x - 5 > 3$ or $7x - 3 > 4$ **13.** $0 < \dfrac{3}{2}x - 3 \le 3$ 13. _____

14. $x - \dfrac{3}{2} \le \dfrac{5}{4}$ and $-\dfrac{2}{3}x - \dfrac{2}{9} < \dfrac{8}{9}$

15. $-3 < -4x + 1 < 17$

14. _____

15. _____

16. $-4 \le \dfrac{4x-3}{3} < 3$

17. $-15 < -3(x+2) \le 1$

16. _____

17. _____

In Problem 18, use the Addition Property and/or Multiplication Properties to find a and b.

18. If $-4 < x < 3$, then $a < 2x - 7 < b$.

18. _____

19. Diastolic Blood Pressure Blood pressure is measured using two numbers. One of the numbers measures diastolic blood pressure. The diastolic blood pressure represents the pressure while the heart is resting between beats. In a healthy person, the diastolic blood pressure should be greater than 60 and less than 90. If we let the variable x represent a person's diastolic blood pressure, express the diastolic blood pressure of a healthy person using compound inequality.

19.

20. Heating Bills For usage above 300 kilowatt-hours, the non-space heat winter energy charge for Illinois Power residential service was $23.12 plus $0.05947 per kilowatt-hour over 300. During one winter, a customer's charge ranged from a low of $50.28 to a high of $121.43. Over what range of values did electric usage vary (in kilowatt-hours)?

20.

Sullivan/Struve, *Intermediate Algebra, 3e*
Copyright © 2014 Pearson Education, Inc.

Five-Minute Warm-Up 2.6
Absolute Value Equations and Inequalities

1. Evaluate each expression.

   (a) $|-12|$ (b) $|0|$ (c) $\left|\dfrac{3}{4}\right|$ (d) $|-5.2|$

   1a. _____

   1b. _____

   1c. _____

   1d. _____

2. Express the distance between the origin, 0, and 45 as an absolute value.

   2. _____

3. Express the distance between the origin, 0, and -12 as an absolute value.

   3. _____

4. Solve each equation.

   (a) $-3x + 7 = -5$ (b) $4(x + 1) = x + 5x - 10$

   4a. _____

   4b. _____

5. Solve each inequality.

   (a) $6x - 10 < 8x + 2$ (b) $\dfrac{1}{2}(3x - 1) \le \dfrac{2}{3}(x + 3)$

   5a. _____

   5b. _____

Guided Practice 2.6
Absolute Value Equations and Inequalities

Objective 1: Solve Absolute Value Equations

1. If a is a positive real number and if u is any algebraic expression, then $|u| = a$ is equivalent to

   _____ or _____.

2. When solving absolute value equations the first step is to _____.

3. Solve the equation $|3x - 1| - 5 = -3$. *(See textbook Example 2)*

Step 1: Isolate the expression containing the absolute value.

$$|3x - 1| - 5 = -3$$

Add 5 to both sides: **(a)** _____

Step 2: Rewrite the absolute value equation as two equations: $u = a$ and $u = -a$, where u is the algebraic expression in the absolute value symbol. Here $u = 3x - 1$ and $a = 2$.

(b) _____ or _____

Step 3: Solve each equation.

(c) $x =$ _____ or $x =$ _____

Step 4: Check. Verify each solution.

Substitute your values for x into the original equation. If the statement is true, then the value is a solution of the absolute value equation. If the statement is false, delete the value from the solution set.

(d) solution set: _____

4. If u and v are any algebraic expressions, then $|u| = |v|$ is equivalent to

   _____ or _____.

Objective 2: Solve Absolute Value Inequalities Involving $<$ or $\leq$

5. If a is a positive real number and if u is any algebraic expression, then

   $|u| < a$ is equivalent to _____.

   $|u| \leq a$ is equivalent to _____.

6. Solve the inequality $|4x - 3| \leq 9$. Write the solution set in interval notation. *(See textbook Example 6)*

Step 1: The inequality is in the form $|u| \leq a$ where $u = 4x - 3$ and $a = 9$. We rewrite the inequality as a compound inequality that does not involve absolute value.

$|4x - 3| \leq 9$

Use the fact that $|u| \leq a$ means $-a \leq u \leq a$:

(a) _____

Step 2: Solve the resulting compound inequality.

Add 3 to all three parts:

(b) _____

Divide all three parts of the inequality by 4:

(c) _____

Graph the solution:

(d)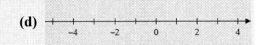

Write the solution in interval notation:

(e) _____

Objective 3: Solve Absolute Value Inequalities Involving $>$ or $\geq$

7. If a is a positive real number and if u is any algebraic expression, then

$|u| > a$ is equivalent to _____ or _____.

$|u| \geq a$ is equivalent to _____ or _____.

8. Solve the inequality $3|8x + 3| > 9$. Write the solution set in interval notation. *(See textbook Example 9)*

Step 1: The inequality is in the form $|u| > a$ where $u = 8x + 3$ and $a = 3$. We rewrite the inequality as a compound inequality that does not involve absolute value.

$3|8x + 3| > 9$

Isolate the absolute value:

(a) _____

Rewrite the inequality:

(b) _____

Step 2: Solve each inequality separately.

(c) _____ or _____

Step 3: Find the union of the solution sets of each inequality.

Graph the solution:

(d)

Write the solution in interval notation:

(e) _____

Sullivan/Struve, *Intermediate Algebra*, 3e
Copyright © 2014 Pearson Education, Inc.

Do the Math Exercises 2.6
Absolute Value Equations and Inequalities

In Problems 1 and 2, solve each absolute value equation.

1. $\left| \dfrac{2x-3}{5} \right| = 2$

2. $3|y-4| + 4 = 16$

1. _____

2. _____

In Problems 3 – 6, solve each absolute value inequality. Express your answer in set-builder notation.

3. $|y+4| < 6$

4. $|-3x+2| - 7 \le -2$

3. _____

4. _____

5. $|x+4| \ge 7$

6. $3|z| + 8 > 2$

5. _____

6. _____

In Problems 7 – 14, solve each absolute value equation or inequality.

7. $|2x+1| = x - 3$

8. $|3 - 5x| < |-7|$

7. _____

8. _____

9. $|3x-4| = -9$

10. $|-9x+2| \ge -1$

9. _____

10. _____

11. $\left|4x+3\right|=1$

12. $\left|4x-3\right|>1$

11. _____

12. _____

13. $\left|7x+5\right|+4<3$

14. $\left|4y+3\right|\geq-3$

13. _____

14. _____

In Problems 15 and 16, write each statement as an absolute value inequality.

15. x differs from -4 by less than 2

16. twice x differs from 7 by more than 3

15. _____

16. _____

17. Gestation Period The length of human pregnancy is about 266 days. It can be shown that a mother whose gestation period x satisfies the inequality $\left|\dfrac{x-266}{16}\right|>1.96$ has an unusual length of pregnancy. Determine the length of pregnancy that would be considered unusual.

17. _____

18. Explain why the solution set of $\left|5x-3\right|>-5$ is the set of all real numbers.

Sullivan/Struve, *Intermediate Algebra*, 3e
Copyright © 2014 Pearson Education, Inc.

Five-Minute Warm-Up 3.1
Systems of Linear Equations in Two Variables

1. Evaluate $5x - 2y$ for $x = 3$, $y = -1$.

1. _____

2. Determine whether the point $\left(8, -\dfrac{4}{3} \right)$ is on the graph of the equation $x - 3y = 12$.

2. _____

3. Graph the linear equation $y = -\dfrac{2}{3}x + 4$.

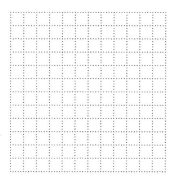

4. Find the equation of the line parallel to $x - y = 2$ containing the point $(-3, -2)$.

4. _____

5. Determine the slope and y-intercept of $x - 3y = -9$.

5.
slope: _____

y-intercept: ___

6. What is the additive inverse of 15?

6. _____

7. Solve: $4x - 2(5x - 1) = -4$

7. _____

Guided Practice 3.1
Systems of Linear Equations in Two Variables

Objective 1: Determine Whether an Ordered Pair Is a Solution of a System of Linear Equations

1. Complete the following chart which describes the solutions to a system of linear equations in two variables.

Number of Solutions	Classification	Graph of the Two Lines
(a) no solution		
(b) infinitely many solutions		
(c) exactly one solution		

Objective 3: Solve a System of Two Linear Equations by Substitution (See textbook Example 4)

2. Solve the following system by substitution: $\begin{cases} 5x + 2y = -5 & (1) \\ 3x - y = -14 & (2) \end{cases}$

Step 1: Solve one of the equations for one of the unknowns. It is easiest to solve equation (2) for y since the coefficient of y is -1.

Equation (2): $3x - y = -14$

Subtract $3x$ from both sides: **(a)** _____

Multiply both sides by -1:   **(b)** _____

Step 2: Substitute your expression for y in equation (1).

Equation (1):   **(c)** _____

Substitute the expression from equation (2) into equation (1):   **(d)** _____

Step 3: Solve the equation for x.

Distribute the 2:   **(e)** _____

Combine like terms: **(f)** _____

Subtract 28 from both sides: **(g)** _____

Divide both sides by 11:   **(h)** _____

Step 4: Substitute the value for x into the equation from Step 1(b) and then solve for y.

(i) $y =$ _____

Step 5: Check your answer in both of the original equations. If both equations yield a true statement, you have the correct answer.

Write the ordered pair that is the solution to the system.   **(j)** _____

Objective 4: Solve a System of Two Linear Equations by Elimination *(See textbook Example 6)*

3. Solve the following system by elimination: $\begin{cases} 2x + y = -4 & (1) \\ 3x + 5y = 29 & (2) \end{cases}$

Step 1: Our first goal is to get the coefficients on one of the variables to be additive inverses. In looking at this system, we can make the coefficients of y be additive inverses by multiplying equation (1) by -5.

Multiply both sides of (1) by -5, use the Distributive Property, and then write the equivalent system of equations.

$\begin{cases} 2x + y = -4 \\ 3x + 5y = 29 \end{cases}$

(a) $\begin{cases} \underline{\hspace{4cm}} \ (1) \\ \underline{\hspace{4cm}} \ (2) \end{cases}$

Step 2: We now add equations (1) and (2) to eliminate the variable y and then solve for x.

Add (1) and (2):　**(b)** _____

Divide both sides by -7:　**(c)** _____

Step 3: Substitute your value for x into either equation (1) or equation (2). We will use equation (1) as it looks like less work.

Equation (1):　　　　　$2x + y = -4$

Substitute your value for x.　**(d)** _____

Solve for y.　**(e)** _____

Step 4: Check your answer in both of the original equations. If both equations yield a true statement, you have the correct answer.

Write the ordered pair that is the solution to the system.　**(f)** _____

Objective 5: Identify Inconsistent Systems

4. Algebraically, what occurs when you solve an inconsistent system of equations? _____

_____

Objective 6: Write the Solution of a System with Dependent Equations

5. Algebraically, what occurs when you solve a dependent system of equations? _____

_____

6. The following system is consistent and dependent. $\begin{cases} -x + 3y = 1 \\ 2x - 6y = -2 \end{cases}$

Express the solution using set-builder notation.　6. _____

Do the Math Exercises 3.1
Systems of Linear Equations in Two Variables

In Problems 1 and 2, determine whether the given ordered pairs are solutions of the system of linear equations.

1a. _____

1b. _____

1. $\begin{cases} x - 2y = -11 \\ 3x + 2y = -1 \end{cases}$

2. $\begin{cases} -3x + y = 5 \\ 6x - 2y = 6 \end{cases}$

    (a) $(-5, 3)$ **(b)** $(-3, 4)$

    (a) $(-2, -1)$ **(b)** $(2, 0)$

2a. _____

2b. _____

In Problems 3 and 4, solve the system of equations by graphing.

3. $\begin{cases} y = -2x + 4 \\ y = 2x - 4 \end{cases}$

4. $\begin{cases} -x + 2y = -9 \\ 2x + y = -2 \end{cases}$

3. _____

4. _____

In Problems 5 and 6, solve the system of equations using substitution.

5. $\begin{cases} y = \dfrac{1}{2}x \\ x - 4y = -4 \end{cases}$

6. $\begin{cases} 3x + 2y = 0 \\ 6x + 2y = 5 \end{cases}$

5. _____

6. _____

In Problems 7 and 8, solve the system of equations using elimination.

7. $\begin{cases} 6x - 4y = 6 \\ -3x + 2y = 3 \end{cases}$

8. $\begin{cases} x + 2y = -\dfrac{8}{3} \\ 3x - 3y = 5 \end{cases}$

7. _____

8. _____

In Problems 9 – 11, solve the system of equations by any method.

9. $\begin{cases} 2x + y = -1 \\ -3x - 2y = 7 \end{cases}$

10. $\begin{cases} y = \dfrac{1}{2}x + 2 \\ x - 2y = -4 \end{cases}$

9. _____

10. _____

11. $\begin{cases} \dfrac{1}{3}x - \dfrac{1}{2}y = -5 \\ -\dfrac{4}{5}x + \dfrac{6}{5}y = 1 \end{cases}$

11. _____

In Problems 12 and 13, use slope-intercept form to determine the number of solutions the system has.

12. $\begin{cases} 4x - 2y = 8 \\ -10x + 5y = 5 \end{cases}$

13. $\begin{cases} 2x - y = -5 \\ -4x + 3y = 9 \end{cases}$

12. _____

13. _____

14. **Rhombus** A rhombus is a parallelogram whose adjacent sides are congruent. Consider the rhombus with vertices $(-1, 3)$, $(3, 6)$, $(3, 1)$, and $(-1, -2)$ to find the following.

   (a) Find the equation of the line for the diagonal through the points $(-1, 3)$ and $(3, 1)$.   14a. _____

   (b) Find the equation of the line for the diagonal through the points $(-1, -2)$ and $(3, 6)$. 14b. _____

   (c) Find the point of intersection of the diagonals.   14c. _____

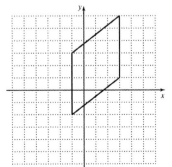

Five-Minute Warm-Up 3.2
Problem Solving: Systems of Two Linear Equations Containing Two Unknowns

In Problems 1 – 4, translate each sentence into a mathematical statement.

1. Twelve more than a number x is 5.

1. _____

2. Six less than a number y is two times y.

2. _____

3. Twice the difference of the length, l, and 6 is equivalent to the width, w.

3. _____

4. The difference of twice the height, h, and 10 is the same as the quotient of 2 and w.

4. _____

5. If a total of \$12,000 is to be invested in stocks and bonds and s represents the amount in stocks, write an algebraic expression for the amount invested in bonds.

5. _____

6. Suppose that you have a credit card balance of \$2,500 and the credit card company charges 18% annual interest on outstanding balances. How much interest will you pay after 1 month?

6. _____

7. Write a linear cost function if the fixed costs are \$4000 and the cost of production is \$42 per unit. Let x represent the number of units produced.

7. _____

Guided Practice 3.2
Problem Solving: Systems of Two Linear Equations Containing Two Unknowns

Objective 1: Model and Solve Direct Translation Problems

1. An adult ticket to the amusement park costs $26, and a child's ticket to the amusement park costs $18.50. A group of 13 friends purchased adult and child's tickets and paid $278. How many tickets for children were purchased? *(See textbook Example 2)*

(a) Sept 1: Identify What do we want to find in the problem? _____

(b) Step 2: Name Let *a* represent the number of adult tickets purchased and *c* represent the number of child's tickets purchased.

    Step 3: Translate Write a system of equations that can be used to solve this problem. Let equation (1) represent the number of tickets purchased and equation (2) represent the total cost of the tickets.

$$\begin{cases} \underline{\hspace{4cm}} & (1) \\ \underline{\hspace{4cm}} & (2) \end{cases}$$

(c) Step 4: Solve   _____

(d) Step 5: Check When you substitute the values you found in Step 4 into your equations from Step 3, does each equation yield a true statement? Is your answer reasonable?

                                               _____

(e) Step 6: Answer _____

Objective 2: Model and Solve Geometry Problems

2. A rectangular parking lot has a perimeter of 125 feet. The length of the parking lot is 10 feet more than the width. What are the dimensions of the parking lot? *(See textbook Example 3)*

(a) Write the formula from Geometry needed to solve this problem.   _____

(b) Write an equation in two variables that relates the length to the width of the parking lot.

                                               _____

(c) Use the equations from (a) and (b), using *l* for the length and *w* for the width of the parking lot, to write a system of two linear equations containing two unknowns that will solve this problem. Be sure your equations are in standard form.

$$\begin{cases} \underline{\hspace{4cm}} & (1) \\ \underline{\hspace{4cm}} & (2) \end{cases}$$

(d) Solve the system.   _____

(e) Check your result and then answer the question. _____

Objective 3: Model and Solve Mixture Problems

3. A backpacker wishes to mix peanuts worth $2 per pound and trail mix worth $5 per pound to make 10 pounds of a cheaper trail mix to take on the family backpacking trip to the Sierra Mountains. If the back packer could find the blend in the grocery store, it would sell for $3.20 per pound. How many pounds of each should be in the mixture if it is worth $3.20 per pound? *(See textbook Example 6)*

(a) Complete the table from the information given, using p for the number of pounds of peanuts and t for the number of pounds of trail mix.

	Number of Pounds	Price per Pound	Total Value
Peanuts			
Trail Mix			
Blend			

(b) Write a system of equations that models this problem.

$$\begin{cases} \underline{\hspace{4cm}} \quad (1) \\ \\ \underline{\hspace{4cm}} \quad (2) \end{cases}$$

(c) Solve the system and answer the question. _____

Objective 4: Model and Solve Uniform Motion Problems

4. Some uniform motion problems involve currents such as wind, water currents, or even standing on a moving walkway in an airport. If r represents the traveling speed without any current and c represents the rate of the current, write an expression for each of the following rates.

(a) speed when traveling against the current   _____

(b) speed when traveling with the current   _____

5. A cyclist can go 36 miles with the wind blowing at her back in 3 hours. On the return trip, after 4 hours, the cyclist sill has 4 miles remaining to return to the starting point. Find the speed of the cyclist and the speed of the wind. *(See textbook Example 7)*

(a) Complete the table from the information given, using c for speed of the cyclist and w for speed of the wind.

	Distance	Rate	Time
With the wind			
Against the wind			

(b) Write a system of equations that models this problem.

$$\begin{cases} \underline{\hspace{4cm}} \quad (1) \\ \\ \underline{\hspace{4cm}} \quad (2) \end{cases}$$

(c) Solve the system and answer the question. _____

Objective 5: Find the Intersection of Two Linear Functions

6. Geometrically, when you set two functions equal to each other and solve, what are you finding?

_____

7. In business, what is a break-even point? _____

Sullivan/Struve, *Intermediate Algebra*, 3e
Copyright © 2014 Pearson Education, Inc.

Do the Math Exercises 3.2
Problem Solving: Systems of Two Linear Equations Containing Two Unknowns

1. The sum of two numbers is 25. The difference of two numbers is 3. Find the numbers.   1. _____

2. The sum of four times a first number and a second number is 68. If the first number is   2. _____
decreased by twice the second number the result is −1. Find the numbers.

3. **Making Change** Johnny has $6.75 in dimes and quarters. He has 8 more dimes than   3. _____
quarters. How many dimes does Johnny have?

4. **Perimeter** The perimeter of a rectangle is 260 centimeters. If the width of the rectangle   4. _____
is 15 centimeters less than the length, what are the dimensions of the rectangle?

5. **Investments** Marge and Homer have $80,000 to invest. Their financial advisor has   5. _____
recommended that they diversify by placing some of the money in stocks and some in
bonds. Based upon current market conditions, he has recommended that three
times the amount in bonds should equal two times the amount invested in stocks. How
much should be invested in bonds?

6. **Candy** A candy store sells chocolate-covered almonds for $6.50 per pound and chocolate-covered peanuts for $4.00 per pound. The manager decides to make a bridge mix that combines the almonds with the peanuts. She wants the bridge mix to sell for $6.00 per pound, and there should be no loss in revenue from selling the bridge mix versus the almonds and peanuts alone. How many pounds of chocolate-covered almonds and chocolate-covered peanuts are required to create 50 pounds of bridge mix?

6. _____

7. **Pharmacy** A doctor's prescription calls for the creation of pills that contain 10 units of vitamin B_{12} and 13 units of vitamin E. Your pharmacy stocks two powders that can be used to make these pills: Powder A contains 20% vitamin B_{12} and 40% vitamin E; Powder B contains 50% vitamin B_{12} and 30% vitamin E. How many units of each powder should be mixed in each pill?

7. _____

8. **Against the Wind** A Piper Arrow can fly 510 miles in 3 hours with a tail wind. Against this same wind, the plane can fly 390 miles in 3 hours. Find the airspeed of the plane. What is the impact of the wind on the plane?

8. _____

9. **Runners** Enrique leaves his house and starts to run at an average speed of 6 miles per hour. Half an hour later, Enrique's younger (and faster) brother leaves the house to catch up to Enrique running at an average speed of 8 miles per hour. How long will it take for Enrique's brother to run half the distance that Enrique has run?

9. _____

Five-Minute Warm-Up 3.3
Systems of Linear Equations in Three Variables

1. Evaluate $2x - 5y - 8z$ for $x = -7, y = 4, z = -3$.   1. _____

In Problems 2 – 4, solve the system of equations using elimination.

2. $\begin{cases} x - y = 10 \\ x + y = -20 \end{cases}$ 2. _____

3. $\begin{cases} 5x - y = 3 \\ -10x + 2y = 2 \end{cases}$ 3. _____

4. $\begin{cases} 4x + y = 3 \\ 8x + 2y = 6 \end{cases}$ 4. _____

Guided Practice 3.3
Systems of Linear Equations in Three Variables

Objective 1: Solve Systems of Three Linear Equations

1. Geometrically, what does an equation in three variables represent? _____

2. A system of three linear equations containing three variables has one of the following possible solutions:

(a) Exactly one solution is a _____ system with _____ equations.

(b) No solution is an _____ system.

(c) Infinitely many solutions is a _____ system with _____ equations.

3. Use the method of elimination to solve the system: $\begin{cases} x + 3y + 3z = 9 & (1) \\ 3x + 5y + 4z = 8 & (2) \\ 5x + 3y + 7z = 9 & (3) \end{cases}$ *(See textbook Example 2)*

Step 1: Our goal is to eliminate the same variable from two of the equations. In looking at the system, we notice that we can use equation (1) to eliminate the variable x from equations (2) and (3). We can do this by multiplying equation (1) by -3 and adding the result to equation (2). The equation that results becomes equation (4). Why do we do this? Because the coefficients on x will be additive inverses and adding the equations eliminates the variable x. We also multiply equation (1) by -5 and add the result to equation (3). The equation that results becomes equation (5).

Multiply equation (1) by -3 : **(a)** _____

Equation (2): **(b)** _____

Add (a) and (b): **(c)** _____ (4)

Multiply equation (1) by -5 : **(d)** _____

Equation (3): **(e)** _____

Add (d) and (e): **(f)** _____ (5)

Step 2: We now concentrate on equations (4) and (5), treating them as a system of two equations containing two variables. It is easiest to eliminate the variable y by multiplying equation (4) by -3 and adding the result to equation (5). This results in an equation in one variable, equation (6).

Multiply equation (4) by -3 : **(g)** _____

Equation (5): **(h)** _____

Add (g) and (h): **(i)** _____ (6)

Step 3: We solve equation (6) for z. On line (i), divide both sides by 7: **(j)** _____

Step 4: Back-substitute your value for z into either equation (4) or equation (5) to solve for y. **(k)** $y =$ _____

Step 5: Back-substitute your values for y and z into one of the equations (1), (2), or (3). Solve for x. **(l)** $x =$ _____

continued next page

Step 6: Check your answer in all three of the original equations. If all equations yield a true statement, you have the correct answer.

Write the ordered triple that is the solution to the system.   **(m)** _____

Objective 2: *Identify Inconsistent Systems*

4. Whenever you solve a system of equations and end up with a false statement such as $0 = -7$, you have an

_____ system. We say that the solution to the system is _____.

Objective 3: *Write the Solution with a System of Dependent Equations*

5. Whenever you solve a system of equations and end up with a true statement such as $3 = 3$ or $0 = 0$, you

have a _____ system.

6. Typically, when writing the solution set, we express the values of x and y in terms of z, although this is not

required. We know that $\begin{cases} 2x - y & = 2 \ \ (1) \\ -x \ \ \ \ + 5z = 3 \ \ (2) \\ \ \ \ \ - y + 10z = 8 \ \ (3) \end{cases}$ is a dependent system. *(See textbook Example 5)*

(a) Solve equation (2) for x: _____

(b) Solve equation (3) for y: _____

(c) Express the solution to the system: $\{(x, y, z) | x =$ _____ $, y =$ _____ $, z$ is any real number$\}$

Objective 4: *Model and Solve Problems Involving Three Linear Equations*

7. Theater Revenues A theater has 600 seats, divided into orchestra, main floor, and balcony seating. Orchestra seats sell for $80, main floor seats for $60, and balcony seats for $25. If all the seats are sold, the total revenue to the theater is $33,500. One evening, all of the orchestra seats were sold, $\frac{3}{5}$ of the main seats were sold and $\frac{4}{5}$ of the balcony seats were sold. The total revenue collected was $24,640. How many are there of each kind of seat? *(See textbook Example 6)*

(a) Write an equation that expresses the total number of seats in the theater if a represents the number of orchestra, b represents the number of main floor, and c represents the number of balcony seats:

_____

(b) Write an equation that calculates the total revenue from all seats: _____

(c) Write an equation that calculates the revenue when a portion of the seats are sold:

_____

Sullivan/Struve, *Intermediate Algebra*, 3e
Copyright © 2014 Pearson Education, Inc.

Do the Math Exercises 3.3
Systems of Linear Equations in Three Variables

Determine whether the given ordered triples are solutions of the system of linear equations.

1.
$$\begin{cases} 2x + y - 2z = 6 \\ -2x + y + 5z = 1 \\ 2x + 3y + z = 13 \end{cases}$$

1a. _____

1b. _____

    (a) $(3, 2, 1)$ **(b)** $(10, -4, 5)$

In Problems 2 – 6, solve each system of three linear equations containing three unknowns.

2.
$$\begin{cases} x + 2y - z = 4 \\ 2x - y + 3z = 8 \\ -2x + 3y - 2z = 10 \end{cases}$$

2. _____

3.
$$\begin{cases} x - y + 3z = 2 \\ -2x + 3y - 8z = -1 \\ 2x - 2y + 4z = 7 \end{cases}$$

3. _____

4.
$$\begin{cases} x \qquad - 3z = -3 \\ \quad 3y + 4z = -5 \\ 3x - 2y \qquad = 6 \end{cases}$$

4. _____

5.
$$\begin{cases} x - y + 2z = 3 \\ 2x + y - 2z = 1 \\ 4x - y + 2z = 0 \end{cases}$$

5. _____

6.
$$\begin{cases} x + y + z = 4 \\ 2x + 3y - z = 8 \\ x + y - z = 3 \end{cases}$$

6. _____

7. Curve Fitting The function $f(x) = ax^2 + bx + c$ is a quadratic function where a, b, and c are constants.

(a) If $f(-1) = 6$, then $6 = a(-1)^2 + b(-1) + c$ or $a - b + c = 6$. Find two additional linear equations if $f(1) = 2$, and $f(2) = 9$.

7a. _____

(b) Use the three linear equations found in part **(a)** to determine a, b, and c.

7b. _____

(c) What is the quadratic function that contains the points $(-1, 6)$, $(1, 2)$, and $(2, 9)$?

7c. _____

8. Nutrition Antonio is on a special diet that requires he consume 1325 calories, 172 grams of carbohydrates, and 63 grams of protein for lunch. He wishes to have a Broccoli and Cheese Baked Potato, Chicken BLT Salad, and a medium Coke. Each Broccoli and Cheese Baked Potato has 480 calories, 80 g of carbohydrates, and 9 g of protein. Each Chicken BLT Salad has 310 calories, 10 g for carbohydrates, and 33 g for protein. Each Coke has 140 calories, 37 g of carbohydrates, and 0 g of protein. How many servings of each does Antonio need?

8. _____

Sullivan/Struve, *Intermediate Algebra,* 3e
Copyright © 2014 Pearson Education, Inc.

Five-Minute Warm-Up 3.4
Using Matrices to Solve Systems

1. Determine the coefficients of $-2x + y - 3z$.

1._____

2. Evaluate $-x - 5y + 11z$ for $x = 2, y = -5, z = -1$.

2._____

In Problems 3 and 4, solve for the indicated variable.

3. $7x - 5y = 10$ for y

4. $\dfrac{3}{2}x + \dfrac{2}{3}y = -2$ for x

3._____

4._____

In Problems 5 and 6, use the Distributive Property to remove the parentheses.

5. $-3(2x - 9y + z)$

6. $-\dfrac{5}{4}(8x - 4y + 12z)$

5._____

6._____

7. If $f(x) = -x^2 - 5x + 7$, find the value of each function.
   (a) $f(4)$ (b) $f(-3)$

7a._____

7b._____

Guided Practice 3.4
Using Matrices to Solve Systems

Objective 1: Write the Augmented Matrix of a System

1. A **matrix** is a rectangular array of numbers, meaning that the order of the numbers in the matrix is relevant. The size of the matrix, called the *dimension*, is denoted as the number of rows by the number of columns. If a matrix has 3 rows and 4 columns, we say the dimension of the matrix is 3×4.

Find the dimension: $\begin{bmatrix} -1 & 8 & 3 & -7 \\ -5 & 0 & 1 & 2 \end{bmatrix}$

1. _____

2. An **augmented matrix** can be used to represent a system of linear equations. Each row is created from one of the equations in the system and each column represents the coefficients of one of the variables. The vertical bar is the equal sign and the last column represents the constants. Be sure each equation is written in

_____ form, filling in the coefficient of any missing variables with _____.

3. Write the system of equations as an augmented matrix. *(See textbook Example 1)*

$$\begin{cases} 2x + y + z = 3 \\ 4y - 7z = -1 \\ x + 3y = 0 \end{cases}$$

3. _____

Objective 2: Write the System from the Augmented Matrix

4. Write the system of linear equations corresponding to the augmented matrix. *(See textbook Example 2)*

$$\begin{bmatrix} 1 & 1 & | & 2 \\ -3 & 1 & | & 10 \end{bmatrix}$$

4. _____

Objective 3: Perform Row Operations on a Matrix

5. There are three basic row operations. These are similar to the types of operations that we used to solve systems of equations earlier in this chapter. List the row operations for matrices.

(a) _____

(b) _____

(c) _____

6. The notation $-3r_1$ means multiply row 1 by -3. The notation $R_2 = r_1 + r_2$ means replace row 2 with the sum of row 1 plus row 2. Perform the following row operations and write the new augmented matrix. *(See textbook Example 3)*

(a) $\begin{bmatrix} 2 & -1 & | & 5 \\ 1 & -7 & | & 4 \end{bmatrix}$ $R_2 = -2r_2 \rightarrow$

(b) $\begin{bmatrix} 2 & -1 & | & 5 \\ 1 & -7 & | & 4 \end{bmatrix}$ $R_1 = r_2 + r_1 \rightarrow$

Objective 4: Solve Systems Using Matrices

7. When is a matrix in **row echelon form**? _____

_____

8. Solve the following system using matrices:
$$\begin{cases} x + y + z = 1 \\ 2x + 2y \quad\;\; = 6 \\ 3x + 4y - z = 13 \end{cases}$$
(See textbook Example 6)

Step 1: Write the augmented matrix of the system. We will use the row operations from Objective 3 to solve the system.

(a) _____

Step 2: We want the entry in row 1, column 1 to be 1. This is already done.

Step 3: We want the entry in row 2, column 1 to be zero. We also want the entry in row 3, column 1 to be a zero. We use row operation #3 to accomplish this. The entries in row 1 remain unchanged.

$R_2 = -2r_1 + r_2 \rightarrow$

$R_3 = -3r_1 + r_3 \rightarrow$

(b) _____

Step 4: We want the entry in row 2, column 2 to be a 1. This is accomplished by interchanging rows 2 and 3, row operation #1.

(c) _____

Step 5: We want the entry in row 3, column 2 to be zero. This is already accomplished.

Step 6: We want the entry in row 3, column 3 to be a 1. We use row operation #2 to accomplish this.

$R_3 = -\dfrac{1}{2}r_3 \rightarrow$

(d) _____

Step 7: The augmented matrix is in row echelon form. Write the system of equations corresponding to the augmented matrix and solve. We know the value of *z*. Substitute into equation (2) and solve for *y*. Then use these values and substitute into equation (1) to solve for *x*.

State the solution as an ordered triple.

(e) _____

Step 8: Check

We leave it to you to verify the solution.

Objective 5: Solve Consistent Systems with Dependent Equations and Solve Inconsistent Systems

9. State the solution to the system represented by the augmented matrix: $\begin{bmatrix} 1 & 4 & | & 2 \\ 0 & 0 & | & 0 \end{bmatrix}$

9. _____

(See textbook Example 7)

10. State the solution to system represented by the augmented matrix: $\begin{bmatrix} 1 & 0 & -1 & | & -2 \\ 0 & 1 & 5 & | & -9 \\ 0 & 0 & 0 & | & -3 \end{bmatrix}$

10. _____

(See textbook Example 8)

Sullivan/Struve, *Intermediate Algebra*, 3e
Copyright © 2014 Pearson Education, Inc.

Do the Math Exercises 3.4
Using Matrices to Solve Systems

●

Write the augmented matrix of the given system of equations.

1. $\begin{cases} 6x + 4y + 2 = 0 \\ -x - y + 1 = 0 \end{cases}$

1. _____

Perform each row operation on the given augmented matrix.

2. $\begin{bmatrix} 1 & -1 & 1 & | & 6 \\ -2 & 1 & -3 & | & 3 \\ 3 & 2 & -2 & | & -5 \end{bmatrix}$

2a. _____

2b. _____

(a) $R_2 = 2r_1 + r_2$ followed by

(b) $R_3 = -3r_1 + r_3$

●

In Problems 3 – 6, solve each system of equations using matrices. If a system has no solution, say that it is inconsistent.

3. $\begin{cases} 5x - 2y = 3 \\ -15x + 6y = -9 \end{cases}$

4. $\begin{cases} 3x + 3y = -1 \\ 2x + y = 1 \end{cases}$

3. _____

4. _____

●

5.
$$\begin{cases} -x + 2y + z = 1 \\ 2x - y + 3z = -3 \\ -x + 5y + 6z = 2 \end{cases}$$

5. _____

6.
$$\begin{cases} 2x - y + 2z = 13 \\ -x + 2y - z = -14 \\ 3x + y - 2z = -13 \end{cases}$$

6. _____

7. Finance Marlon has $12,000 to invest. He decides to place some of the money into a savings account paying 2% annual interest, some in Treasury bonds paying 4% annual interest and some in a mutual fund paying 9% annual interest. Marlon would like to earn $440 per year in income. In addition, Marlon wants his investment in the savings account to be $4,000 more than the amount in Treasury bonds. How much should Marlon invest in each investment category?

7. _____

Sullivan/Struve, *Intermediate Algebra*, 3e
Copyright © 2014 Pearson Education, Inc.

Five-Minute Warm-Up 3.5
Determinants and Cramer's Rule

1. Evaluate: $-3 \cdot 5 - 2 \cdot (-7)$

1. _____

2. Simplify each expression.

(a) $\dfrac{13}{0}$

(b) $\dfrac{0}{45}$

2a. _____

2b. _____

3. Simplify: $\dfrac{-12}{-10}$

3. _____

4. Evaluate $x - 2y + z$ for $x = -\dfrac{3}{2}$, $y = \dfrac{2}{5}$, $z = \dfrac{3}{4}$.

4. _____

In Problems 5 and 6, solve each equation.

5. $-8 - 3x = 1$

6. $\dfrac{9}{7}z - 4 = 32$

5. _____

6. _____

7. Solve: $\begin{cases} -6 - 2(3x - 6y) = 0 \\ 6 - 12(x - 2y) = 0 \end{cases}$

7. _____

Guided Practice 3.5
Determinants and Cramer's Rule

Objective 1: Evaluate the Determinant of a 2×2 Matrix

1. The notation $\begin{vmatrix} a & b \\ c & d \end{vmatrix}$ denotes the **determinant** for the matrix $\begin{bmatrix} a & b \\ c & d \end{bmatrix}$. We use the definition

$\begin{vmatrix} a & b \\ c & d \end{vmatrix} = ad - bc$ to evaluate the determinant and find a single number representing the array.

Evaluate each determinant. *(See textbook Example 1)*

(a) $\begin{vmatrix} 3 & 8 \\ -1 & 2 \end{vmatrix}$ **(b)** $\begin{vmatrix} -7 & 2 \\ 9 & 1 \end{vmatrix}$

1a. _____

1b. _____

Objective 2: Use Cramer's Rule to Solve a System of Two Equations

2. If the number of unknowns is the same as the number of equations in a system, we can use Cramer's Rule to solve the system. We first set up several different determinants.

(a) D is a determinant in which each entry is a _____ of the variables in the system.

(b) D_x is a determinant in which the first column (coefficients of x) is replaced by _____.

(c) D_y is a determinant in which the _____ column (coefficients of ___) is replaced by the constants on the right side of the equal sign.

3. Given the linear equations $-5y + 3x = 9$ and $2 - x - 2y = 0$, write the system with each equation expressed in standard form.

3. _____

4. Using your system above, determine each of the following. *(See textbook Example 2)*

     (a) $D =$ **(b)** $D_x =$ **(c)** $D_y =$

5. According to Cramer's Rule, the solution to the system of equations can be found evaluating the determinants D, D_x, and D_y and simplifying the following ratios.

     (a) $x = \dfrac{?}{?}$ **(b)** $y = \dfrac{?}{?}$

Objective 3: Evaluate the Determinant of a 3×3 Matrix

6. We cannot use the process from Objective 1 to find the value of a 3×3 determinant. Do you see why? Instead we use a process called **expansion by minors**. Although it is possible to expand by any row or column, typically we use the first row entries to be the coefficients of the corresponding minors. Be careful with the operations between the terms as they alternate in the expansion and can change when expanding by a different row or column. *(See textbook Example 3)*

Set up, but do not evaluate, the expansion by row 1: $\begin{vmatrix} 1 & -1 & 2 \\ 5 & -3 & 3 \\ 1 & 4 & -2 \end{vmatrix}$

_____

Objective 4: Use Cramer's Rule to Solve a System of Three Equations *(See textbook Example 5)*

7. Solve the following system using Cramer's Rule: $\begin{cases} x + 2y - z = -3 \\ 2x - 4y + z = -7 \\ -2x + 2y - 3z = 4 \end{cases}$

Step 1: Find the determinant of the coefficients of the variables, D.

Write the determinant, D: **(a)** _____

Evaluate D: **(b)** _____

Step 2: Because $D \neq 0$, we continue by writing and evaluating each of the determinants D_x, D_y, and D_z.

Write and evaluate D_x: **(c)** _____

Write and evaluate D_y: **(d)** _____

Write and evaluate D_z: **(e)** _____

Step 3: Solve for each of the variables.

(f) $x = \dfrac{D_x}{D} = \dfrac{?}{?}$

$= \underline{\quad}$

(g) $y = \dfrac{?}{?} = \dfrac{?}{?}$

$= \underline{\quad}$

(h) $z = \dfrac{?}{?} = \dfrac{?}{?}$

$= \underline{\quad}$

Step 4: Check you answer.

We leave it to you to verify the solution.

State your solution as an ordered triple. **(i)** _____

8. If $D = 0$ Cramer's Rules does not apply. If at least one of the determinants D_x, D_y, or D_z is different from

0, then the system is_____ and the solution set is _____ .

9. If $D = 0$ Cramer's Rule does not apply. If all of the determinants D_x, D_y, and D_z equal 0, then the system is

_____ and _____ and there are _____ solutions.

Sullivan/Struve, *Intermediate Algebra*, 3e
Copyright © 2014 Pearson Education, Inc.

Do the Math Exercises 3.5
Determinants and Cramer's Rule

In Problems 1 – 3, find the value of each determinant.

1. $\begin{vmatrix} 5 & 3 \\ 2 & 4 \end{vmatrix}$

2. $\begin{vmatrix} -8 & 5 \\ -4 & 3 \end{vmatrix}$

1. _____

2. _____

3. $\begin{vmatrix} -2 & 1 & 6 \\ -3 & 2 & 5 \\ 1 & 0 & -2 \end{vmatrix}$

3. _____

In Problems 4 – 7, solve each system of equations using Cramer's Rule, if possible.

4. $\begin{cases} 2x + 4y = -6 \\ 3x + 2y = 7 \end{cases}$

5. $\begin{cases} 3x - 6y - 2 = 0 \\ x + 2y - 4 = 0 \end{cases}$

4. _____

5. _____

6. $\begin{cases} x + y - z = 6 \\ x + 2y + z = 6 \\ -x - y + 2z = -7 \end{cases}$

6. _____

7. $\begin{cases} x - 2y - z = 1 \\ 2x + 2y + z = 3 \\ 6x + 6y + 3z = 6 \end{cases}$

7. _____

8. *Solve for x:* $\begin{vmatrix} -2 & x \\ 3 & 4 \end{vmatrix} = 1$

8. _____

9. Geometry: Area of a Triangle Given the points $A = (-1, -1)$, $B = (3, 2)$, and $C = (0, 6)$, find the area of the triangle ABC.

9. _____

Sullivan/Struve, *Intermediate Algebra*, 3e
Copyright © 2014 Pearson Education, Inc.

Five-Minute Warm-Up 3.6
Systems of Linear Inequalities

1. Determine whether $x = -5$ satisfies the inequality $-3x - 10 \geq 5$.

1.

2. Determine whether the ordered pair $(4, -2)$ satisfies the linear inequality $2x - 5y < 10$.

2.

3. Solve the inequality: $-\dfrac{1}{2}(3x + 1) > \dfrac{2}{3}(3 - x)$

3.

4. Graph the linear inequality $3x - 4y > 0$.

Guided Practice 3.6
Systems of Linear Inequalities

Objective 2: Graph a System of Linear Inequalities

1. The only way to show the solution of a system of linear inequalities is by _____.

2. To graph the inequality $x - 8y > 4$, we use a _____ line as the boundry.
   (Refer to Section 1.8 in textbook)

3. To graph the inequality $3x + 2y \geq -6$, we use a _____ line as the boundry.

4. To graph a linear inequality, we first graph the equation to determine the boundary line. This line divides the plane into two half-planes. To decide which half-plane to shade, we use a test point such as $(0, 0)$, provided that this point does not lie on the line. If the test point satisfies the inequality, we shade the half-plane that contains the point. If the test point does not satisfy the inequality, we shade

_____.

5. When graphing a system of linear inequalities, we are looking for the ordered pairs that satisfy both inequalties simultaneously. Therefore, the solution of the system is the _____ of the graphs of the linear inequalties.

6. Graph the system: $\begin{cases} 2x + 3y > -3 \\ -3x + y \leq 2 \end{cases}$ *(See textbook Example 2)*

Step 1: Graph the inequality $2x + 3y > -3$ on the graph at the right.

Step 2: Graph the inequality $-3x + y \leq 2$ on the graph at the right.

Step 3: Looking at what you graphed in Steps 1 and 2, determine where the shaded regions overlap.

Objective 3: Solve Problems Involving Systems of Linear Inequalities

7. Party Planning Alexis and Sarah are planning a barbeque for their friends. They plan to serve grilled fish and carne asada and want to spend at most $40 on the meat. The fish sells for $8 per pound and the carne asada is $5 per pound. Since most of their friends do not eat red meat, they plan to buy at least twice as much fish as carne asada. Let *x* represent the amount of carne asada purchased and *y* represent the amount of fish purchased. *(See textbook Example 6)*

(a) Write an inequality that describes how much Alexis and Sarah will spend on meat.   7a. _____

(b) Write an inequality that describes the amount of carne asada that will be purchased   7b. _____
relative to the amount of fish that will be purchased.

(c) Since Alexis and Sarah will purchase a positive quantity of meat, write the two   7c. _____
inequalities that describe this constraint.

_____

(d) Graph the system of inequalities.

(e) Identify the vertices of the polygon.   7e. _____

_____

_____

Sullivan/Struve, *Intermediate Algebra*, 3e
Copyright © 2014 Pearson Education, Inc.

Do the Math Exercises 3.6
Systems of Linear Inequalities

In Problems 1 and 2, determine which of the points, if any, satisfies the system.

1. $\begin{cases} x + y \geq 2 \\ -3x + y \leq 10 \end{cases}$

2. $\begin{cases} 5x + 2y < 10 \\ 4x - 3y < 24 \end{cases}$

1. _____

2. _____

(a) $(-3, 6)$ **(b)** $(4, 1)$

(a) $(1, 3)$ **(b)** $(1, 1)$

In Problems 3 – 6, graph each system of inequalities.

3. $\begin{cases} x + y \geq 2 \\ -3x + y \leq 10 \end{cases}$

4. $\begin{cases} -x + \dfrac{1}{3}y < 3 \\ \dfrac{4}{3}x + y \geq 4 \end{cases}$

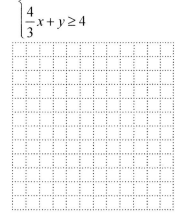

5. $\begin{cases} 4x + 3y > -9 \\ -8x - 6y > 12 \end{cases}$

6. $\begin{cases} y \leq 4 \\ x \geq -1 \end{cases}$

Graph the system of linear inequalities. Tell whether the graph is bounded or unbounded and label the corner points.

7.
$$\begin{cases} x + y \geq 8 \\ x + 3y \geq 12 \\ x \geq 0 \\ y \geq 0 \end{cases}$$

8. **Mixing Nuts** You've Got to Be Nuts is a store that specializes in selling nuts. The owner finds that she has excess inventory of 100 pounds (1600 ounces) of cashews and 120 pounds (1920 ounces) of peanuts. She decides to make two types of 1 pound nut mixes from the excess inventory. A premium mix will contain 12 ounces of cashews and 4 ounces of peanuts while the standard mix will contain 6 ounces of cashews and 6 ounces of peanuts.

(a) Use x to denote the number of premium mixes and y denote the number of standard mixes. Write a system of linear inequalities that describe the possible number of each kind of mix.

8a. _____

(b) Find the corner points of the graph.

8b. _____

Five-Minute Warm-Up 4.1
Adding and Subtracting Polynomials

In Problems 1 and 2, identify the coefficient.

1. $-x^2$ **2.** $5y^4$ 1. _____

                                                                                  2. _____

3. Combine like terms: $3x^2 + 8x - 6x + x^2$ 3. _____

4. Use the Distributive Property to remove the parentheses: $-8(6x + 2)$ 4. _____

5. Simplify: $8p - 3(p - 4)$ 5. _____

6. Combine like terms: $x^2y - 2xy^2 + 3xy^2 - x^2y$ 6. _____

In Problems 7 and 8, find each function value.

7. $f(x) = -11x + 5;\ f(-2)$ **8.** $g(x) = -x^2 + 3x - 1;\ g(-4)$ 7. _____

                                                                                  8. _____

Name: Date:
Instructor: Section:

Guided Practice 4.1
Adding and Subtracting Polynomials

Objective 1: Define Monomial and Determine the Coefficient and Degree of a Monomial

1. For our study of polynomials, we begin with some definitions regarding monomials.
(See textbook Example 1)

(a) In your own words, what is a *monomial*? _____

(b) What is the coefficient of a monomial? _____

(c) How do you determine the degree of a monomial in one variable? _____

(d) How do you determine the degree of a monomial in more than one variable? *(See textbook Example 3)*

_____

Objective 2: Define Polynomial and Determine the Degree of a Polynomial

2. Next, we move on to some definitions used with polynomials. *(See textbook Example 4)*

(a) In your own words, define *polynomial*. _____

(b) When is a polynomial in *standard form*? _____

(c) How do you determine the degree of a polynomial? _____

3. Some polynomials have special names. Always simplify the polynomial first, if possible, before determining if the polynomial has one of the following specific names:

(a) a polynomial with exactly one term is a _____

(b) a polynomial with exactly two terms is a _____

(c) a polynomial with exactly three terms is a _____

(d) a polynomial with more than three terms is simply called a _____

Objective 3: Simplify Polynomials by Combining Like Terms

4. In your own words, define *like terms*. _____

5. To add two polynomials, we need to combine the like terms of the polynomials. The parentheses are included to indicate the first polynomial added to a second polynomial.

Simplify using horizontal addition: $\left(9x^2 - x + 5\right) + \left(3x^2 + 1\right)$ *(See textbook Example 6)*

Step 1: Remove parenthesis. **(a)** _____

Step 2: Rearrange terms. **(b)** _____

Step 3: Use the distributive property in reverse to group the coefficients of the like terms. **(c)** _____

Step 4: Simplify. **(d)** _____

6. Simplify by subtracting the polynomials: $\left(6z^3 + 2z^2 - 5\right) - \left(-3z^3 + 9z^2 - z + 1\right)$

   (See textbook Example 8)

Step 1: Distribute the -1. **(a)** _____

Step 2: Rearrange terms. **(b)** _____

Step 3: Combine like terms. **(c)** _____

Objective 5: Add and Subtract Polynomial Functions

7. The *sum* of two functions f and g is defined by $(f + g)(x) = f(x) + g(x)$. This is read "the function f plus g of x is equal to f of x plus g of x". *(See textbook Example 11)*

     Let f and g be two polynomial functions defined as
$$f(x) = 6x^2 - 5x + 1 \text{ and } g(x) = -4x^2 + 5x - 7, \text{ find}$$

(a) $f(-4)$ **(b)** $g(-4)$ **(c)** $f(-4) + g(-4)$ 7a. _____

7b. _____

7c. _____

(d) $(f + g)(x)$ **(e)** $(f + g)(-4)$ 7d. _____

7e. _____

Sullivan/Struve, *Intermediate Algebra*, 3e
Copyright © 2014 Pearson Education, Inc.

Do the Math Exercises 4.1
Adding and Subtracting Polynomials

In Problems 1 – 3, determine the coefficient and degree of each monomial.

1. $5x^4$ **2.** $-12xy$ **3.** -7

1. _____

2. _____

3. _____

In Problems 4 – 7, determine whether the algebraic expression is a polynomial (Yes or No).
If it is a polynomial, determine the degree and state if it is a monomial, binomial or trinomial.
If it is a polynomial with more than 3 terms, identify the expression as polynomial.

4. _____

4. $\dfrac{1}{x}$ **5.** -12

5. _____

6. _____

6. $4y^{-2} + 6y - 1$ **7.** $4mn^3 - 2m^2n^3 + mn^8$

7. _____

In Problems 8 – 12, simplify each polynomial by adding or subtracting, as indicated.
Express your answer as a single polynomial in standard form.

8. _____

8. $10y^4 - 6y^4$ **9.** $\left(8 - t^3\right) - \left(1 + 3t + 3t^2 + t^3\right)$

9. _____

10. $\left(-5xy^2 + 3xy - 9y^2\right) - \left(5xy^2 + 7xy - 8y^2\right)$ 10. _____

11. $\left(\dfrac{3}{4}y^3 - \dfrac{1}{8}y + \dfrac{2}{3}\right) + \left(\dfrac{1}{2}y^3 + \dfrac{5}{12}y - \dfrac{5}{6}\right)$ 11. _____

12. $\left(2w^3 - w^2 + 6w - 5\right) + \left(-3w^3 + 5w^2 + 9\right)$ 12. _____

In Problems 13 – 15, given $f(x) = -2x^3 + 3x - 1,$ *find each the following values.*

13. $f(0)$ 14. $f(2)$

13. _____

14. _____ ●

15. $f(-3)$

15. _____

In Problems 16 and 17, for the given functions f and g, find the following functions.

16. $f(x) = 4x + 3;\ g(x) = 2x - 3;\ find\ (f+g)(x).$

16. _____

17. $f(x) = 3x^2 + x + 2;\ g(x) = x^2 - 3x - 1;\ find\ (f-g)(1).$

17. _____

In Problems 18 and 19, perform the indicated operations.

18. Add $2x^3 - 3x^2 - 5x + 7$ to $x^3 + 3x^2 - 6x - 4.$

18. _____ ●

19. Subtract $2q^3 - 3q^2 + 7q - 2$ from $-5q^3 + q^2 + 2q - 1.$

19. _____

20. The polynomial function $I(a) = -54.42a^2 + 4853.38a - 46,106.66$ can be used to approximate the average per-capita income, I, of a U.S. resident in 2009, where a is the age of the individual.

   (a) Use the function to estimate the average income of a 20 year old in 2009.

   20a. _____

   (b) Use the function to estimate the average income of a 55 year old in 2009.

   20b. _____

Five-Minute Warm-Up 4.2
Multiplying Polynomials

In Problems 1 – 6, simplify each expression.

1. $x^8 \cdot x$

2. $7y^3 \cdot 3y^4$

1. _____

2. _____

3. $(4z)^3$

4. $(-2a^2)^4$

3. _____

4. _____

5. $\left(\dfrac{5}{2}x\right)^2$

6. $\left(\dfrac{4xy^2}{3xy}\right)^2 \cdot \left(\dfrac{x^3y^4}{2xy^5}\right)^3$

5. _____

6. _____

7. Use the Distributive Property to remove the parentheses: $7(3a - 2b)$

7. _____

8. Simplify: $\dfrac{2}{3}\left(\dfrac{6x}{5} + \dfrac{9}{8}\right)$

8. _____

Guided Practice 4.2
Multiplying Polynomials

Objective 1: Multiply a Monomial and a Polynomial

1. To multiply a monomial and a polynomial we use _____.

2. It is important that you are familiar with the Laws of Exponents. If you need more practice multiplying monomials or using the Product Rule for Exponents, review Getting Ready for Chapter 4: Laws of Exponents and Scientific Notation.

Multiply and simplify: $2ab^2\left(-a^2 - 7ab + 3b^2\right)$ 2. _____

(See textbook Example 2)

Objective 2: Multiply Two Binomials

3. When multiplying two binomials, you have two of options. These two methods are:

(a) _____   **(b)** _____

Practice both methods and then select the method that you prefer for multiplying the binomials in the exercise set and the remainder of the course.

4. Use the distributive property to find the product: $(9x - 2)(3x + 5)$ 4. _____

(See textbook Example 3)

5. Use FOIL to find the product: $(4a - 7b)(3a + b)$ 5. _____

(See textbook Example 4)

Objective 3: Multiply Two Polynomials

6. Find the product: $(3x - 1)(2x^2 + x - 5)$ *(See textbook Example 5)*

$$(3x - 1)(2x^2 + x - 5)$$

Step 1: Distribute the binomial $3x - 1$ to each term
in the trinomial.   **(a)** _____

Step 2: Distribute.   **(b)** _____

Step 3: Combine like terms.   **(c)** _____

Objective 4: Multiply Special Products

7. Some products occur frequently and are given special names. One special product is called the difference of squares. It states that $(A - B)(A + B) =$ _____.
(See textbook Example 6)

8. Another special product is called the square of binomials. It states that **(a)** $(A + B)^2 =$ _____ and **(b)** $(A - B)^2 =$ _____. *(See textbook Example 8)*

Objective 5: Multiply Polynomial Functions

9. The *product* of two functions f and g is defined by $(f \cdot g)(x) = f(x) \cdot g(x)$. This is read "the function f times g of x is equal to f of x times g of x". *(See textbook Example 10)*

Let f and g be two polynomial functions defined as
$$f(x) = 7x - 2 \text{ and } g(x) = x^2 - 5, \text{ find}$$

(a) $f(-1)$ **(b)** $g(-1)$ **(c)** $f(-1) \cdot g(-1)$

9a. _____

9b. _____

9c. _____

(d) $(f \cdot g)(x)$ **(e)** $(f \cdot g)(-1)$

9d. _____

9e. _____

Sullivan/Struve, *Intermediate Algebra*, 3e
Copyright © 2014 Pearson Education, Inc.

Do the Math Exercises 4.2
Multiplying Polynomials

In Problems 1 – 3, find the product.

1. $(9a^3b^2)(-3a^2b^5)$

2. $\left(\dfrac{12}{5}x^2y\right)\left(\dfrac{15}{4}x^4y^3\right)$

1. _____

2. _____

3. $-3mn^3(4m^2 - mn + 5n^2)$

3. _____

In Problems 4 – 12, find the product of the polynomials.

4. $(z - 8)(z + 3)$

5. $(2 - 7y)(5 + 2y)$

4. _____

5. _____

6. $\left(\dfrac{3}{2}y + 4\right)\left(\dfrac{4}{3}y - 1\right)$

7. $(3m - 5n)(m + 2n)$

6. _____

7. _____

8. $(3p^2 - 5p + 3)(7p - 2)$

9. $(xy - 2)(x^2 + 2xy + 4y^2)$

8. _____

9. _____

10. $(7p - 3q)^2$

11. $(m^2 - 2n^3)(m^2 + 2n^3)$

10. _____

11. _____

12. $[(m + 4) - n]^2$

12. _____

In Problems 13 and 14, given the functions $f(x) = (x + 5)$ *and* $g(x) = x^2 - 2x + 3,$ *find*

13. $(f \cdot g)(x)$ **14.** $(f \cdot g)(3)$

13. _____

14 _____

In Problems 15 and 16, given the function $f(x) = -2x^2 + x - 5,$ *find*

15. $f(x + 2)$ **16.** $f(x + h) - f(x)$

15. _____

16. _____

In Problems 17 – 23, simplify the expression.

17. $-3x(x - 3)^2$ **18.** $(4x + 3)(3x - 7)$

17. _____

18. _____

19. $(2y + 3)(4y^2 - 6y + 9)$ **20.** $\left(3x + \dfrac{1}{3}\right)^2$

19. _____

20. _____

21. $(z - 3)^3$ **22.** $(2a + b - 5)(4a - 2b + 1)$

21. _____

22. _____

23. $(a + 2)(a - 2)(a^2 - 4) - (a + 3)(a^2 - 3)$

23. _____

Sullivan/Struve, *Intermediate Algebra*, 3e
Copyright © 2014 Pearson Education, Inc.

Five-Minute Warm-Up 4.3
Dividing Polynomials; Synthetic Division

In Problems 1 – 6, simplify each expression.

1. $\dfrac{r^7}{r^2}$

2. $\dfrac{63x^8}{27x^4}$

1. _____

2. _____

3. $25z^0$

4. $\dfrac{8a^4b}{16ab^3}$

3. _____

4. _____

5. $\left(\dfrac{9s}{2t^5}\right)^2$

6. $\left(\dfrac{xy^2z}{2xz^2}\right)^3 \cdot \left(\dfrac{4xyz}{y^3}\right)^2$

5. _____

6. _____

7. Add: $\dfrac{4}{9} + \dfrac{8}{9}$

7. _____

Guided Practice 4.3
Dividing Polynomials; Synthetic Division

Objective 2: Divide Polynomials Using Long Division

1. Find the quotient and remainder when $\left(x^3 - 2x^2 + x + 6\right)$ is divided by $\left(x + 1\right)$.

(See textbook Example 3)

Step 1: Divide the leading term of the dividend, x^3, by the leading term of the divisor, x. Enter the result over the term x^3.

(a) $\dfrac{x^3}{x} = $ _____

Step 2: Multiply ☐ by $x + 1$. Be sure to vertically align like terms.

(b) $\begin{array}{r} \square \\ x+1\overline{)x^3 - 2x^2 + x + 6} \\ \hline \end{array}$

Step 3: Subtract your product from the dividend.

(c) $\begin{array}{r} x^2 \\ x+1\overline{)x^3 - 2x^2 + x + 6} \\ \hline \end{array}$

Subtract and continue with Steps 4 and 5:

Step 4: Repeat Steps 1 – 3 treating $-3x^2 + x + 6$ as the dividend and dividing x into $-3x^2$ to obtain the next term in the quotient.

Step 5: Repeat Steps 1 – 3 treating $4x + 6$ as the dividend and dividing x into $4x$ to obtain the next term in the quotient.

When the degree of the remainder is less than the degree of the divisor, you are finished dividing.

Express the answer as the

Quotient $+ \dfrac{\text{Remainder}}{\text{Divisor}}$.

(d) _____

Step 6: Check Verify that (Quotient)(Divisor) + Remainder = Dividend

We leave it to you to verify the solution.

Dividing polynomials using long division can be difficult. Practice dividing an integer by an integer as shown in Example 2 before continuing to divide polynomials using long division. Did you notice that the steps are exactly the same? In Step 3, we subtracted the two polynomials and started the process over in Step 4. It is also possible to subtract only the like terms and then bring down the next before starting the division process again. Try it both ways. Which way do you prefer?

Objective 3: Divide Polynomials Using Synthetic Division

2. When using synthetic division to divide two polynomials, it is essential that the dividend be in

_____ form and, if any of the powers of the variable are missing, fill in a _____ coefficient for that term.

3. Synthetic division can be used only when the divisor is of the form _____ or _____ .

4. Use synthetic division to find the quotient and remainder when $\left(18 + x^4 - 9x^2 + 3x^3\right)$ is divided by $(x - 3)$. *(See textbook Example 6)*

Step 1: Write the dividend in descending powers of x. Then copy
the coefficients of the dividend. Remember to insert a 0 for any
missing power of x.

(a)_____

Step 2: Insert the division symbol. Rewrite the divisor in the form
$x - c$ and insert the value of c to the left of the division symbol.

(b)_____

Step 3: Bring the 1 down 2 rows and enter it in Row 3.

(c)_____

Step 4: Multiply the latest entry in Row 3 by 3 and place the result
in Row 2, one column to the right.

(d) _____

Step 5: Add the entry in Row 2 to the entry above it in Row 1. Enter
the sum in Row 3.

(e)_____

Step 6: Repeat steps 4 and 5 until no more entries are available in
Row 1.

(f)_____

Step 7: The final entry in Row 3, 99, is the remainder; the other
entries in Row 3, 1, 6, 9, and 27, are the coefficients of the quotient,
in descending order of degree. The quotient is the polynomial whose
degree is one less than the degree of the dividend. State the quotient
and the remainder.

(g)_____

Step 8: Check: (Quotient)(Divisor) + Remainder = Dividend. Write
your final answer as $\text{quotient} + \dfrac{\text{remainder}}{\text{divisor}}$

(h) _____

Objective 5: Use the Remainder and Factor Theorems

5. The Remainder Theorem Let f be a polynomial function. If $f(x)$ is divided by $x - c$, then the

remainder is _____ .

6. The Factor Theorem Let f be a polynomial function. Then $x - c$ is a factor of $f(x)$ if and only if

_____ .

Sullivan/Struve, *Intermediate Algebra*, 3e
Copyright © 2014 Pearson Education, Inc.

Do the Math Exercises 4.3
Dividing Polynomials; Synthetic Division

In Problems 1 – 3, divide and simplify.

1. $\dfrac{6z^3 + 9z^2}{3z^2}$

2. $\dfrac{3z^4 + 12z^2}{6z^3}$

1. _____

2. _____

3. $\dfrac{2x^2y^3 - 9xy^3 + 16x^2y}{2x^2y^2}$

3. _____

In Problems 4 – 9, divide using long division.

4. $\dfrac{x^2 - 4x - 21}{x + 3}$

5. $\dfrac{4x^2 - 17x - 33}{4x + 7}$

4. _____

5. _____

6. $\dfrac{x^3 + x^2 - 22x - 40}{x + 2}$

7. $\dfrac{a^3 - 49a + 120}{a + 8}$

6. _____

7. _____

8. $\dfrac{x^3 - 5x^2 - 2x + 10}{x^2 - 2}$

9. $\dfrac{2k^3 + 10k^2 - 6k - 8}{2k^2 - 3}$

8. _____

9. _____

In Problems 10 – 12, divide using synthetic division.

10. $\dfrac{x^2 + 2x - 17}{x - 4}$

10._____

11. $\dfrac{x^3 - 13x - 17}{x + 3}$

11._____

12. $\dfrac{a^4 - 65a^2 + 55}{a - 8}$

12._____

In Problems 13 and 14, given $f(x) = 3x^2 - 6x + 5$ *and* $g(x) = 2x + 1$, *find*

13. $\left(\dfrac{f}{g}\right)(x)$

14. $\left(\dfrac{f}{g}\right)(2)$

13._____

14._____

In Problem 15, use the Remainder Theorem to find the remainder.

15. $f(x) = 3x^3 + 2x^2 - 5$ is divided by $x + 3$

15._____

In Problem 16, use the Factor Theorem to determine whether $x - c$ *is a factor of the given function for the given value of c.*

16. $f(x) = x^2 + 5x + 6;\ c = 3$

16._____

17. Area The area of a rectangle is $\left(15x^2 + x - 2\right)$ square feet. If the width of the rectangle is $(3x - 1)$ feet, find the length.

17._____

18. Volume The volume of a box is $\left(4x^3 + 7x^2 - 165x - 126\right)$ cubic feet. Find the height if the width is $(4x + 3)$ feet and the length is $(x - 6)$ feet.

18._____

Sullivan/Struve, *Intermediate Algebra*, 3e
Copyright © 2014 Pearson Education, Inc.

Five-Minute Warm-Up 4.4
Greatest Common Factor; Factoring by Grouping

In Problems 1 – 4, write each number as a product of prime factors.

1. 12

2. 18

1. _____

2. _____

3. 72

4. 125

3. _____

4. _____

5. Use the Distributive Property to remove the parentheses: $-8(3x - 2)$

5. _____

6. Multiply and simplify: $7u^3(u^2 + 4u - 1)$

6. _____

7. List the prime numbers less than 30.

7. _____

8. Given $12 \cdot 8 = 96$

   (a) list the factors

   (b) identify the product

8. _____

9. Identify the missing factor: $6x^3 \cdot ? = 18x^5 y$

9. _____

Name: Date:
Instructor: Section:

Guided Practice 4.4
Greatest Common Factor; Factoring by Grouping

Objective 1: Factor Out the Greatest Common Factor

1. Factor out the greatest common factor: $6m^4n^2 + 18m^3n^4 - 22m^2n^5$ *(See textbook Example 2)*

Step 1: Find the GCF.

(a) GCF = _____

Step 2: Rewrite each term as the product of the GCF and remaining factor.

(b) $6m^4n^2 + 18m^3n^4 - 22m^2n^5 = $ _____

Step 3: Factor out the GCF.

(c) $6m^4n^2 + 18m^3n^4 - 22m^2n^5 = $ _____

Step 4: Check   Distribute to verify that the factorization is correct.

2. Factor out the greatest common binomial factor: $7(a-1)^2 - 14(a-1)$ *(See textbook Example 4)*

2. _____

Objective 2: Factor by Grouping

3. Factor by grouping is commonly used when a polynomial has _____ terms.

4. *True or False* You may need to rearrange the terms in order to be able to identity a common binomial factor.

4. _____

5. Factor by grouping: $2x^2 - 4x + 3xy - 6y$ *(See textbook Example 5)*

Step 1: Group terms with common factors. In this problem the first two terms have a common factor and the last two terms have a common factor.

(a) common factor of $2x^2 - 4x$: _____

(b) common factor of $3xy - 6y$: _____

Step 2: In each grouping, factor out the common factor.

(c) $2x^2 - 4x + 3xy - 6y =$ _____

Step 3: Factor out the common factor that remains.

(d) $2x^2 - 4x + 3xy - 6y =$ _____

Step 4: Check

Multiply to verify that the factorization is correct.

Sullivan/Struve, *Intermediate Algebra,* 3e
Copyright © 2014 Pearson Education, Inc.

Do the Math Exercises 4.4
Greatest Common Factor; Factoring by Grouping

In Problems 1 – 7, factor out the greatest common factor.

1. $8z + 48$

2. $-4b + 32$

1. _____

2. _____

3. $12a^2 + 45a$

4. $-6q^3 + 36q^2 - 48q$

3. _____

4. _____

5. $-18b^3 + 10b^2 + 6b$

6. $6z(5z + 3) + 5(5z + 3)$

5. _____

6. _____

7. $8a^4b^2 + 12a^3b^3 - 36ab^4$

7. _____

In Problems 8 – 13, factor by grouping.

8. $8x - 8y + bx - by$

9. $3y^3 + 9y^2 - 5y - 15$

8. _____

9._ _____

10. $3a^2 - 15a - 9a + 45$

11. $2y^3 + 14y^2 - 4y^2 - 28y$

10. _____

11. _____

12. $(x + 5)(x - 3) - (x - 1)(x - 3)$

13. $c^3 - c^2 + 5c - 5$

12. _____

13. _____

14. Surface Area The surface area of a cylindrical can whose radius is r inches and height
is 4 inches is given by $S = 2\pi r^2 + 8\pi r$. Express the surface area in factored form.

15. _____

15. Summer Clearance Suppose that an electronics store decides to sell last year's model
televisions for a 20% discount.

    (a) Let x represent the original price of the television. Write an algebraic expression
representing the selling price of the television.

15a. _____

    (b) After 1 month, the manager of the store discounts the TVs by another 15%.
Write an algebraic expression representing the sale price of the televisions in
terms of x, the original selling price.

15b. _____

    (c) Write the algebraic expression in simplified form.

15c. _____

    (d) If the original price of the television was \$650, what is the sale price after the
second discount?

15d. _____

16. The Better Deal Which is the better deal: (a) receiving a 30% discount or (b) receiving
a 15% discount and then another 15% discount after the first 15% discount was applied?
Make up an example that demonstrates how you came to this conclusion.

16. _____

Sullivan/Struve, *Intermediate Algebra*, 3e
Copyright © 2014 Pearson Education, Inc.

Five-Minute Warm-Up 4.5
Factoring Trinomials

1. Determine the coefficients of $3x^2 - 4x - 7$.

1. _____

In Problems 2 and 3, find (a) the sum of the numbers and (b) the product of the numbers.

2. -9 and -4 **3.** -12 and 3

2a. _____

2b. _____

3a. _____

3b. _____

In Problems 4 – 7, find two integers with the following properties.

4. sum of 3 and product of -28

4. _____

5. sum of -15 and product of 54

5. _____

6. sum of 16 and product of 48

6. _____

7. sum of 18 and product of -63

7. _____

8. List the factors of 24 whose sum is 10.

8. _____

9. List the factors of -36 whose sum is -9

9. _____

Guided Practice 4.5
Factoring Trinomials

● **Objective 1: Factor Trinomials of the Form $x^2 + bx + c$**

Step 1: Find the pair of integers whose product is c and whose sum is b. That is, determine m and n such that $mn = c$ and $m + n = b$.

Step 2: Write $x^2 + bx + c = (x + m)(x + n)$.

Step 3: Check your work by multiplying out the factored form.

1. Factor: $y^2 + 9y + 18$ *(See textbook Example 1)*

Step 1: We are looking for factors of $c = 18$ whose sum is $b = 9$. We begin by listing all factors of 18 and computing the sum of these factors.

(a)

Factors whose product is 18	1,18	2,9	3,6	−1, −18	−2, −9	−3, −6
Sum of factors						

Which two factors add to 9 and multiply to 18?   **(b)** _____

Step 2: We write the trinomial in the form $(y + m)(y + n)$.

(c) $y^2 + 9y + 18 = $ _____

Step 3: Check We multiply to verify our solution.

We leave it to you to verify the factorization.

2. How can you tell if a polynomial is *prime* (not factorable)? *(See textbook Example 4)* _____

_____

In addition to the textbook example, if the trinomial is of the form $ax^2 + bx + c$, the polynomial is only factorable when the expression $b^2 - 4ac$ results in a perfect square. Evaluating $b^2 - 4ac$ for the trinomial $y^2 + 9y + 18$ results in 9 so the trinomial is factorable. Evaluating $b^2 - 4ac$ for the trinomial $y^2 + 4y + 12$ results in −32 so the trinomial is not factorable.

Objective 2: Factor Trinomials of the Form $ax^2 + bx + c, a \neq 1$

3. There are two possible methods for factoring trinomials of this type. Name the two methods.

(a) _____

(b) _____

● After Problem 6, we will present a third method that is not in the textbook.

4. Factor by grouping: $3x^2 - 13x + 12$ *(See textbook Example 7)*

Step 1: Find the value of *ac*. **(a)** Identify the coefficients: $a = $ _____, $c = $ _____

(b) The value of $a \cdot c = $ _____

Step 2: We want to determine the integers whose product is 36 and whose sum is -13. We know the factors must have the same sign in order to have a positive product. For the sum to be negative, both factors must be negative.

(c)

Factors whose product is 36	$-1, -36$	$-2, -18$	$-3, -12$	$-4, -9$	$-6, -6$
Sum of factors					

(d) Which two factors multiply to 36 and add to -13? _____

Step 3: Write
$ax^2 + bx + c = ax^2 + mx + nx + c$

(e) $3x^2 - 13x + 12 = $ _____

Step 4: Factor the expression in Step 3 by grouping.

(f) _____

_____

Step 5: Check

FOIL to verify that the factorization is correct.

5. Factor by trial and error: $48x^2 - 4xy - 30y^2$
   (See textbook Examples 9 – 12)

5. _____

Objective 3: Factor Trinomials Using Substitution

6. Factor: $12p^4 - p^2 - 1$ *(See textbook Example 13)*

Step 1: Rewrite the trinomial in the form $au^2 + bu + c$.

(a) Let $u = $ _____

Substitute.

(b) $12p^4 - p^2 - 1 = $ _____

Step 2: Factor the trinomial from part (b).

(c) _____

Step 3: Rewrite the factored expression from part (c), in *p*.

(d) _____

Alternative Approach to Factor $ax^2 + bx + c$, $a \ne 1$, Using Synthetic Factoring

Another method for factoring trinomials of the form $ax^2 + bx + c$, $a \ne 1$, is synthetic factoring. When using this method, it is very important to first factor out any common factors.

Sullivan/Struve, *Intermediate Algebra*, 3e
Copyright © 2014 Pearson Education, Inc.

EXAMPLE :   How to Factor $ax^2 + bx + c$, $a \neq 1$, **by Synthetic Factoring**

Factor: $6x^2 + 11x + 3$

First, notice that $6x^2 + 11x + 3$ has no common factors. In this trinomial, $a = 6$, $b = 11$, and $c = 3$.

Step 1: Find the value of ac	The value of $a \cdot c$ is $6 \cdot 3 = 18$.
Step 2: Find the pair of integers, m and n, whose product is ac and whose sum is b.	We want to find integers whose product is 18 and whose sum is 11. Because both 18 and 11 are positive, we only list the positive factors of 18. 	Factors whose product is 18	1, 18	2, 9	3, 6	 	Sum of factors	19	11	9	
Step 3: Write two fractions in the form $\dfrac{a}{m}$ and $\dfrac{a}{n}$.	We have $m = 2$, $n = 9$, and $a = 6$, so $\dfrac{a}{m} = \dfrac{6}{2}$ and $\dfrac{a}{n} = \dfrac{6}{9}$.
Step 4: Write each fraction in lowest terms so that $\dfrac{a}{m} = \dfrac{p}{q}$ and $\dfrac{a}{n} = \dfrac{r}{s}$.	$\dfrac{a}{m} = \dfrac{6}{2} = \dfrac{3}{1} = \dfrac{p}{q}$ $\qquad$ $\dfrac{a}{n} = \dfrac{6}{9} = \dfrac{2}{3} = \dfrac{r}{s}$
Step 5: Write the factors $(px + q)(rx + s)$.	From Step 4, we have $p = 3$, $q = 1$, $r = 2$, and $s = 3$, so $$6x^2 + 11x + 3 = (3x + 1)(2x + 3)$$
Step 6: Check by multiplying the factors.	Multiply $(3x + 1)(2x + 3)$ to verify that the factorization is correct.

We summarize below the steps used in the example above.

Factoring $ax^2 + bx + c$, $a \neq 1$, **By Synthetic Factoring, Where** a, b, **and** c **Have No Common Factors**

Step 1: Find the value of ac
Step 2: Find the pair of integers, m and n, whose product is ac and whose sum is b.
Step 3: Write two fractions in the form $\dfrac{a}{m}$ and $\dfrac{a}{n}$.
Step 4: Write each fraction in lowest terms so that $\dfrac{a}{m} = \dfrac{p}{q}$ and $\dfrac{a}{n} = \dfrac{r}{s}$.
Step 5: Write the factors $(px + q)(rx + s)$.
Step 6: Check by multiplying the factors.

---

7. Factor using synthetic factoring: $-16x^2 - 4x + 6$

  (a) Factor out the common factor: _____

(b) Determine each of the following using part (a): $a =$ _____ , $b =$ _____ , $c =$ _____ , $a \cdot c =$ _____

Factors of −24								
Sum								

(c) Which two factors multiply to $a \cdot c$ and add to b? _____

(d) $\dfrac{a}{m} = \dfrac{?}{?}$ $\dfrac{p}{q} = \dfrac{?}{?}$ $p =$ _____ and $q =$ _____

(e) $\dfrac{a}{n} = \dfrac{?}{?}$ $\dfrac{r}{s} = \dfrac{?}{?}$ $r =$ _____ and $s =$ _____

(f) Write the factors $(px + q)(rx + s)$ _____

(g) Write the factored form of $-16x^2 - 4x + 6$: _____

Work Smart

Let's compare factoring by grouping and synthetic factoring by factoring $3x^2 + 10x + 8$. First, notice there is no GCF in the trinomial.

Grouping

Step 1: For $3x^2 + 10x + 8$, $a = 3$, $b = 10$, and $c = 8$, so $a \cdot c = 3 \cdot 8 = 24$.

Step 2: The factors of 24 whose sum is $b = 10$ are 6 and 4.

Step 3: Write $3x^2 + 10x + 8$ as
$$3x^2 + 10x + 8 = 3x^2 + 6x + 4x + 10$$

Step 4: Factor $3x^2 + 6x + 4x + 8$ by grouping.

$$3x^2 + 6x + 4x + 8 = (3x^2 + 6x) + (4x + 8)$$
$$= 3x(x + 2) + 4(x + 2)$$
$$= (x + 2)(3x + 4)$$

Synthetic Factoring

Step 1: For $3x^2 + 10x + 8$, $a = 3$, $b = 10$, and $c = 8$, so $a \cdot c = 3 \cdot 8 = 24$.

Step 2: The factors of 24 whose sum is $b = 10$ are 6 and 4.

Step 3: $\dfrac{a}{m} = \dfrac{3}{6}$ and $\dfrac{a}{n} = \dfrac{3}{4}$

Step 4: $\dfrac{a}{m} = \dfrac{3}{6} = \dfrac{1}{2} = \dfrac{p}{q}$ and $\dfrac{a}{n} = \dfrac{3}{4} = \dfrac{r}{s}$

Step 5: $(px + q)(rx + s) = (x + 2)(3x + 4)$, so

$$3x^2 + 10x + 8 = (x + 2)(3x + 4)$$

Both methods give the same result. Many people prefer trial and error while others prefer a more directed process such as grouping or synthetic factoring. You might know a process which has not been discussed here. Which method do you prefer and why?

---

8. Factor: $21n^2 - 18n^3 + 9n$ 8. _____

Sullivan/Struve, *Intermediate Algebra*, 3e
Copyright © 2014 Pearson Education, Inc.

Do the Math Exercises 4.5
Factoring Trinomials

In Problems 1 – 22, factor each polynomial completely. If the polynomial cannot be factored, say it is prime.

1. $z^2 + 3z - 28$ **2.** $q^2 + 2q - 80$ 1. _____

2. _____

3. $-p^2 + 3p + 54$ **4.** $m^2 + 7mn + 10n^2$ 3. _____

4. _____

5. $-4s^2 - 32s - 48$ **6.** $6x^2 - 37x + 6$ 5. _____

6. _____

7. $12r^2 + 11r - 15$ **8.** $20r^2 + 23r + 6$ 7. _____

8. _____

9. $3m^2 + 7mn - 6n^2$ **10.** $4x^3 - 52x^2 + 144x$ 9. _____

10. _____

11. $54x^3y + 33x^2y - 72xy$ **12.** $y^4 + 5y^2 + 6$ 11. _____

12. _____

13. $r^2s^2 + 8rs - 48$

14. $3(z + 3)^2 + 14(z + 3) + 8$

13. _____

14. _____

15. $24m^2 + 58mn + 9n^2$

16. $r^6 - 6r^3 + 8$

15. _____

16. _____

17. $p^2 - 14pq + 45q^2$

18. $9(a + 2)^2 - 10(a + 2) + 1$

17. _____

18. _____

19. $a^2 + a - 6$

20. $24y^2 + 39y - 18$

19. _____

20. _____

21. $-24m^3n - 18m^2n^2 + 27mn^3$

22. $54x^3y + 33x^2y - 72xy$

21. _____

22. _____

Sullivan/Struve, *Intermediate Algebra,* 3e
Copyright © 2014 Pearson Education, Inc.

Five-Minute Warm-Up 4.6
Factoring Special Products

1. List the perfect squares that are less than 100. That is, find $1^2, 2^2, 3^2$ etc.

_____

2. List the perfect cubes that are less than 100. That is, find $1^3, 2^3, 3^3$ etc.

_____

In Problems 3 – 6, evaluate each expression.

3. $\left(\dfrac{5}{3}\right)^2$

4. -8^2

3. _____

4. _____

5. $\left(-\dfrac{4}{3}\right)^2$

6. $\left(-\dfrac{5}{2}\right)^3$

5. _____

6. _____

In Problems 7 and 8, simplify each expression.

7. $\left(4x^2 y\right)^3$

8. $\left(\dfrac{3}{2}ab^3\right)^2$

7. _____

8. _____

9. Multiply: $\left(3x - 2\right)^2$

9. _____

Guided Practice 4.6
Factoring Special Products

●   **Objective 1: Factor Perfect Square Trinomials**

1. Find the product:

(a) $(A + B)^2 = $ _____

(b) $(A - B)^2 = $ _____

2. Factor the perfect square trinomial $p^2 + 18p + 81$. *(See textbook Example 1)*

Step 1: Write the trinomial in the form $A^2 + 2AB + B^2$.   **(a)** Let $A = $ _____

                                                                (b) Let $B = $ _____

$$p^2 + 18p + 81 = \textbf{(c)}\underline{\hspace{5cm}}$$

Step 2: Factor using $A^2 + 2AB + B^2 = (A + B)^2$.   **(d)** _____

3. What has to be added to $4x^2 + 36x$ to make it a perfect square trinomial? _____

●   **Objective 2: Factor the Difference of Two Squares**

4. Find the product: $(A - B)(A + B) = $ _____

5. Factor each difference of two squares completely. *(See textbook Example 2)*

(a) $x^2 - 64$ **(b)** $25m^6 - 36n^4$ 5a. _____

                                                                           5b. _____

6. When a polynomial has four or more terms, try factoring by grouping. In Section 4.4, we grouped two terms in each group. Here we will group three terms to form a perfect square trinomial (which ultimately factors into the difference of two perfect squares). Factor: $x^2 + 4x + 4 - 4y^2$. *(See textbook Example 3)*

Step 1: The first three terms form a perfect square trinomial.

Group the first three terms. $x^2 + 4x + 4 - 4y^2 = $ **(a)** _____

Step 2: Rewrite the expression as the difference of two squares. Use $A^2 + 2AB + B^2 = (A + B)^2$ to factor the perfect square trinomial.   **(b)** _____

●   **Step 3:** Use $A^2 - B^2 = (A + B)(A - B)$ **(c)** _____

Step 4: Check. Distribute to verify that factorization is correct.

Objective 3: Factor the Sum or Difference of Two Cubes

7. Find the product:

(a) $(A + B)(A^2 - AB + B^2) =$ _____

(b) $(A - B)(A^2 + AB + B^2) =$ _____

(c) $(A + B)^3 =$ _____

8. Factor the sum of two cubes: $p^3 + 64$. *(See textbook Example 4)*

Step 1: Write $p^3 + 64$ in the form $A^3 + B^3$. **(a)** Let $A =$ _____

(b) Let $B =$ _____

(c) $A^3 + B^3 =$ _____

Step 2: Factor using $A^3 + B^3 = (A + B)(A^2 - AB + B^2)$. **(d)** $p^3 + 64$ _____

9. Factor completely: $24x^4 + 375x$. *(See textbook Example 5)*

Step 1: Factor out the greatest common factor. $24x^4 + 375x =$ **(a)** _____

Step 2: Rewrite the sum of two cubes in the form $A^3 + B^3$. **(b)** _____

Step 3: Factor completely using $A^3 + B^3 = (A + B)(A^2 - AB + B^2)$. **(c)** _____

Name: Date:
Instructor: Section:

Do the Math Exercises 4.6
Factoring Special Products

In Problems 1 – 4, factor each perfect square trinomial completely.

1. $9z^2 - 6z + 1$

2. $36b^2 + 84b + 49$

1. _____

2. _____

3. $4a^2 + 20ab + 25b^2$

4. $b^4 + 8b^2 + 16$

3. _____

4. _____

In Problems 5 – 7, factor the difference of two squares completely.

5. $81 - a^2$

6. $x^4 - 9y^2$

5. _____

6. _____

7. $36x^2z - 64y^2z$

7. _____

In Problems 8 – 12, factor the sum or difference of two cubes completely.

8. $z^3 + 64$

9. $216 - n^3$

8. _____

9. _____

10. $16m^3 + 54n^3$

11. $(2z + 3)^3 + 27z^3$

10. _____

11. _____

12. $m^9 + n^{12}$

12. _____

In Problems 13 – 24, factor each polynomial completely.

13. $9a^2 - b^2$　　　　　　　　　　**14.** $64x^3 - 125$

13. _____

14. _____

15. $3m^4 - 81mn^3$　　　　　　　　　**16.** $p^4 - 18p^2 + 81$

15. _____

16. _____

17. $9m^2n^2 - 30mn + 25$　　　　　　**18.** $p^2 + 8p + 16 - q^2$

17. _____

18. _____

19. $-5a^3 - 40$　　　　　　　　　　**20.** $36m^2 + 12mn + n^2 - 81$

19. _____

20. _____

21. $y^2 - 3y + 9$　　　　　　　　　**22.** $p^2 - 4p + 4 - q^2$

21. _____

22. _____

23. $x^2 + 0.6x + 0.09$　　　　　　　**24.** $\dfrac{a^2}{36} - \dfrac{b^2}{49}$

23. _____

24. _____

Sullivan/Struve, *Intermediate Algebra*, 3e
Copyright © 2014 Pearson Education, Inc.

Five-Minute Warm-Up 4.7
Factoring: A General Strategy

In Problems 1 – 4, find each product.

1. $\left(-4x^2y^4\right)\left(2xy^3\right)$

2. $\dfrac{4}{3}ab\left(\dfrac{9}{2}a^2b - \dfrac{3}{4}a^3 + \dfrac{15}{8}b\right)$

1. _____

2. _____

3. $\left(4x + 3y\right)\left(5x - 2y\right)$

4. $\left(a + 4b\right)^2$

3. _____

4. _____

In Problems 5 and 6, factor completely.

5. $-27a^3 + 9a^2 - 18a$

6. $4p^2 - 8pq + 3p - 6q$

5. _____

6. _____

Guided Practice 4.7
Factoring: A General Strategy

Objective 1: Factor Polynomials Completely

1. Review the **Steps for Factoring** listed at the beginning of Section 4.7. Step 1 should always be

_____, if possible.

2. Factor: $-12x^3 - 20x^2y + 8xy^2$ *(See textbook Example 1)*

Step 1: Factor out the greatest common factor (GCF), if any exists.	Determine the GCF:	**(a)** GCF = _____
	Factor out the GCF:	**(b)** $-12x^3 - 20x^2y + 8xy^2 =$ _____
Step 2: Count the number of terms.	How many terms are in the polynomial in parentheses?	**(c)** _____
Step 3: We concentrate on the trinomial in parentheses. It is not a perfect square trinomial. Use either factoring by grouping or the trial and error method to factor the trinomial whose leading coefficient is not 1.	Factor the trinomial:	**(d)** _____
Step 4: Check		We leave it to you to multiply and then distribute to verify the factors are correct.

3. Factor: $64a^2 - 25b^4$ *(See textbook Example 2)*

Step 1: Factor out the greatest common factor (GCF), if any exists.	There is no GCF.	
Step 2: Count the number of terms.	How many terms are in the polynomial?	**(a)** _____
Step 3: Because the first term $64a^2 = (8a)^2$, and the second term, $25b^4 = (5b^2)^2$, are both perfect squares, we have the difference of two squares.	Factor the binomial:	**(b)** _____
Step 4: Check		We leave it to you to multiply to verify the factors are correct.

Objective 2: Write Polynomial Functions in Factored Form

4. Write the polynomial function $f(x) = -x^3 - 2x^2 + 5x + 10$ in factored form. *(See textbook Example 8)*

Step 1: Factor out the greatest common factor (GCF), if any exists. **(a)**_____

Step 2: Attempt to factor by grouping. Group the first two terms and the last two terms. $f(x) = -x^3 - 2x^2 + 5x + 10 =$ **(b)**_____

Step 3: In each grouping, factor out the common factor. **(c)**_____

Step 4: Factor out the common binomial factor. **(d)**_____

Step 5: Check. We leave it to you to verify that the factorization is complete.

5. What does factored completely mean?

Sullivan/Struve, *Intermediate Algebra*, 3e
Copyright © 2014 Pearson Education, Inc.

Do the Math Exercises 4.7
Factoring: A General Strategy

In Problems 1 – 20, factor each polynomial completely.

1. $3x^2 + 6x - 105$

2. $-5a^2 + 80$

1. _____

2. _____

3. $8m^2 - 42m + 49$

4. $54p^6 - 2q^3$

3. _____

4. _____

5. $-4c^3 + 16c^2 - 28c$

6. $18t^2 - 9t - 20$

5. _____

6. _____

7. $12p^2 + 50q^2$

8. $16w^4 - 1$

7. _____

8. _____

9. $4w^2 - 3w - 6$

10. $20p^3q - 2p^2q - 4pq$

9. _____

10. _____

11. $54p^5 + 16p^2q^3$

12. $4z^4 - 25$

11. _____

12. _____

13. $4b^4 + 4b^2 - 15$

14. $(3x + 5)^2 + 4(3x + 5) - 21$

13. _____

14. _____

15. $a^2 + 12a + 36 - 4b^2$

16. $w^6 + 4w^3 - 5$

15. _____

16. _____

17. $q^6 + 1$

18. $-2y^3 - 4y^2 + 32y + 64$

17. _____

18. _____

19. $-5z - 20z^3$

20. $18h^5 + 154h^3 - 72h$

19. _____

20. _____

In Problems 21 – 24, factor each polynomial function.

21. $f(x) = x^2 + 3x - 40$

22. $H(p) = 5p^2 + 28p - 12$

21. _____

22. _____

23. $g(x) = -100x^2 + 36$

24. $G(x) = 2x^3 - x^2 - 18x + 9$

23. _____

24. _____

Sullivan/Struve, *Intermediate Algebra*, 3e
Copyright © 2014 Pearson Education, Inc.

Five-Minute Warm-Up 4.8
Polynomial Equations

In Problems 1 and 2, solve each equation.

1. $2x - 3 = 0$

2. $5(2x - 1) + 7 = 0$

1. _____

2. _____

3. Evaluate $3x^2 - 5x + 6$ when **(a)** $x = 9$ and **(b)** $x = -\dfrac{1}{2}$.

3a. _____

3b. _____

4. If $f(x) = -2x + 3$, solve $f(x) = -3$. What point is on the graph of *f*?

4. _____

_____

5. If $f(x) = \dfrac{7}{5}(2x + 5)$, find $f(-5)$. What point is on the graph of *f*?

5. _____

_____

6. Find the zero of $f(x) = \dfrac{8}{3}x + 4$.

6. _____

Guided Practice 4.8
Polynomial Equations

Objective 1: Solve Polynomial Equations Using the Zero-Product Property

1. State the **Zero-Product Property**. _____

_____

_____

2. A second-degree equation is also called a _____

3. What does it mean to write a quadratic equation in *standard form*? _____

_____

4. Solve: $2x^2 - 3x = 2$ *(See textbook Example 2)*

Step 1: Write the quadratic equation in standard form, $ax^2 + bx + c = 0$.

$$2x^2 - 3x = 2$$

Subtract 2 from both sides: **(a)** _____

Step 2: Factor the expression on the left side of the equation.   **(b)** _____

Step 3: Set each factor to 0.   **(c)** _____   **(d)** _____

Step 4: Solve each first-degree **(e)** _____   **(f)** _____
equation.

Step 5: Check Substitute your values into the original equation.

We leave it to you to verify the solutions.

Write the solution set: **(g)** _____

5. Solve: $(x - 2)(x - 3) = 56$ *(See textbook Example 3)*

(a) To solve this equation, the first step is _____

(b) Next, write the quadratic equation in _____

(c) Solve and state the solution set. _____

6. Solve: $p^3 + 2p^2 - 9p = 18$ *(See textbook Example 4)*

Step 1: We put the equation in standard form by subtracting 18 from both sides of the equation.

$$p^3 + 2p^2 - 9p = 18$$

(a) _____

Step 2: Factor the expression on the left side of the equation. Because there are four terms, we factor by grouping.

(b) _____

Step 3: Set each factor to 0.   (c) _____   (d) _____   (e) _____

Step 4: Solve each first-degree equation. (f) _____   (g) _____   (h) _____

Step 5: Check Substitute your values into the original equation.

We leave it to you to verify the solutions.

Write the solution set: (i) _____

Objective 2: Solve Equations Involving Polynomial Functions

7. Suppose $f(x) = 3x^2 - 13x - 10$. *(See textbook Example 6)*

(a) Write an equation to find the zeros of f.

(a)_____

(b) Use the equation from part (a) to find the zeros of f.

(b)_____

(c) What are the x-intercepts of the graph of the function?

(c)_____

Objective 3: Model and Solve Problems Involving Polynomials

8. The length and width of two sides of a rectangle are consecutive odd integers. The area of the rectangle is 255 square centimeters. Find the dimensions of the rectangle.

Step 1: Identify This is a geometry problem involving the area of a rectangle. It also involves consecutive odd integers.

(a) If n represents one of the odd integers, express the next consecutive odd integer: _____

(b) In general, what formula do we use to calculate the area of a rectangle? _____

(c) Step 2: Name Let _____ represent the width of the rectangle and _____ represent the length.

(d) Step 3: Translate Write an equation that will model the area of the rectangle. _____

(e) Step 4: Solve the equation from step 3. _____

(f) Step 5: Check Is your answer reasonable? _____ Does it meet the necessary conditions? _____

(g) Step 6: Answer the question. _____

Sullivan/Struve, *Intermediate Algebra*, 3e
Copyright © 2014 Pearson Education, Inc.

Do the Math Exercises 4.8
Polynomial Equations

In Problems 1 − 14, solve each equation.

1. $2x(3x + 4) = 0$ **2.** $3a(a − 9)(a + 11) = 0$

1. _____

2. _____

3. $5c^2 + 15c = 0$ **4.** $x^2 + 3x − 40 = 0$

3. _____

4. _____

5. $a^2 + 12a + 36 = 0$ **6.** $4c^2 + 6 = 25c$

5. _____

6. _____

7. $6z^2 + 17z = −5$ **8.** $−6n^2 − 9n + 60 = 0$

7. _____

8. _____

9. $\dfrac{2}{3}x^2 + \dfrac{7}{3}x = 5$ **10.** $y(y + 4) = 45$

9. _____

10. _____

11. $w^3 + 5w^2 − 16w − 80 = 0$ **12.** $−24a^3 + 27a = 18a^2$

11. _____

12. _____

13. $(x + 7)(x - 3) = 11$ **14.** $-7z^2 + 42z = 0$ 13. _____

14. _____

In Problems 15 – 16, suppose that $f(x) = x^2 + 5x + 3$. *Find the values of* x *such that*

15. $f(x) = 3$ **16.** $f(x) = 17$ 15. _____

16. _____

17. Find the zeros of the function $h(x) = 8x^2 - 18x - 35$.   17. _____

18. Area The base of a triangle is 4 meters shorter than its height. What are the height and   18. _____
base of the triangle if its area is 48 square meters?

19. Landscape Design Robert Boehm just designed a cloister (a rectangular garden   19. _____
surrounded by a covered walkway on all four sides). The outside dimensions of the
garden are 12 feet by 8 feet, and the area of the garden and the walkway together are 252
square feet. What is the width of the walkway?

20. Making a Box A box is to be made from a rectangular piece of corrugated cardboard,   20. _____
where the length is 8 inches more than the width, by cutting a square piece 3 inches on
side from each corner. The volume of the box is to be 315 cubic inches. Find the
dimensions of the rectangular piece of cardboard.

Five-Minute Warm-Up 5.1
Multiplying and Dividing Rational Expressions

In Problems 1 – 2, factor completely.

1. $x^2 - 3x - 10$

1._____

2. $6z^2 + z - 2$

2._____

3. Solve: $x^2 + 9x + 20 = 0$

3._____

4. Determine the reciprocal of $\dfrac{9}{2}$.

4._____

5. Determine which of the following are in the domain of the variable x for the expression
$\dfrac{2x}{x^2 - x - 6}$.

5._____

    (a) $x = -3$ **(b)** $x = 0$ **(c)** $x = -2$

In Problems 6 – 7, perform the indicated operation. Be sure to express the answer in lowest terms.

6. $\dfrac{9}{25} \cdot \dfrac{10}{9}$ **7.** $\dfrac{9}{4} \div \dfrac{75}{15}$

6._____

7._____

Guided Practice 5.1
Multiplying and Dividing Rational Expressions

Objective 1: Determine the Domain of a Rational Expression

1. To determine the domain of a rational expression we exclude values that cause _____.

2. For each of the following rational expressions, determine the domain. *(See textbook Example 1)*

    (a) $\dfrac{x-4}{x+2}$

    (b) $\dfrac{a^2-2a-3}{a^2-9a+18}$

    2a. _____

    2b. _____

Objective 2: Simplify Rational Expressions

3. To simplify a rational expression, we use the Reduction Property. Remember to always factor first and then divide out common factors. *(See textbook Examples 2 and 3)*

Simplify the rational expression $\dfrac{x^2-8x-20}{2x^2+3x-2}$.

$$\dfrac{x^2-8x-20}{2x^2+3x-2}=$$

Step 1: Factor the numerator and denominator.   (a)_____

Step 2: Divide out common factors using the Reduction Property.   (b) _____

Objective 3: Multiply Rational Expressions *(See textbook Example 4)*

4. Multiply $\dfrac{x^2-4x}{x^2-4}\cdot\dfrac{x^2-x-6}{x-4}$. Simplify the product.

Step 1: Completely factor each polynomial in the numerator and denominator.

$$\dfrac{x^2-4x}{x^2-4}\cdot\dfrac{x^2-x-6}{x-4}=$$ (a) _____ · _____

Step 2: Multiply using $\dfrac{a}{b}\bullet\dfrac{c}{d}=\dfrac{ac}{bd}$

$$= $$ (b) _____ · _____

Step 3: Divide out common factors in the numerator and denominator.

$$= $$ (c) _____

Express the answer as a simplified rational expression in factored form.

(d) _____

Objective 4: Divide Rational Expressions

5. To divide two rational expressions, we rewrite the division problem as an equivalent multiplication problem and then follow the steps for multiplying rational expressions. If x, y, p and q are rational expressions and $y \neq 0$, $p \neq 0$, $q \neq 0$, then

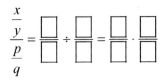

$$\frac{\dfrac{x}{y}}{\dfrac{p}{q}} = \frac{\square}{\square} \div \frac{\square}{\square} = \frac{\square}{\square} \cdot \frac{\square}{\square}$$

6. Divide the following rational expression. Simplify the quotient, if possible. *(See textbook Example 6)*

$$\frac{\dfrac{27xy^2}{60x^2y^4}}{\dfrac{36x}{24y^5}} =$$

Step 1: Rewrite the division problem as a multiplication problem.

(a)_____

Step 2: Multiply.

(b)_____

Step 3: Divide out common factors and simplify.

(c)_____

Objective 5: Work with Rational Functions

7. A **rational function** is a function of the form $R(x) = \dfrac{p(x)}{q(x)}$ where p and q are polynomial functions and q is not the zero polynomial. The domain consists of all real numbers except those for which _____.

8. Find the domain of $R(x) = \dfrac{4x}{2x^2 + 6x - 80}$. *(See textbook Example 7)*

8. _____

Do the Math Exercises 5.1
Multiplying and Dividing Rational Expressions

In Problems 1 – 3, state the domain of each rational function.

1. $\dfrac{4}{x-7}$

2. $\dfrac{x-2}{x^2+4}$

1. _____

2. _____

3. $\dfrac{x+5}{x^2+8x+16}$

3. _____

In Problems 4 – 7, simplify each rational expression.

4. $\dfrac{x^2-3x}{x^2-9}$

5. $\dfrac{w^2+5w-14}{w^2+6w-16}$

4. _____

5. _____

6. $\dfrac{x^2-xy-6y^2}{x^2-4y^2}$

7. $\dfrac{v^3+3v^2-5v-15}{v^2+6v+9}$

6. _____

7. _____

In Problems 8 – 13, multiply each rational expression. Simplify the product, if possible.

8. $\dfrac{5x^2}{x+3}\cdot\dfrac{x^2+7x+12}{20x}$

9. $\dfrac{3x^2+14x-5}{x^2+x-30}\cdot\dfrac{x^2-2x-15}{3x^2+8x-3}$

8. _____

9. _____

10. $\dfrac{2y^2-5y-12}{2y^2-y-6}\cdot\dfrac{4y^2-5y-6}{4-y}$

11. $\dfrac{a^2+2ab+b^2}{3a+3b}\cdot\dfrac{b-a}{a^2-b^2}$

10. _____

11. _____

12. $\dfrac{x^3-27}{2x^2+5x-25} \cdot \dfrac{x^2+2x-15}{x^3+3x^2+9x}$

13. $\dfrac{5m-5}{m^2+6m} \cdot \dfrac{m^2+2m-24}{m^2+3m-4}$

12. _____

13. _____

In Problems 14 – 17, divide each rational expression. Simplify the quotient, if possible.

14. $\dfrac{\dfrac{x-2}{3x}}{\dfrac{5x-10}{x}}$

15. $\dfrac{\dfrac{9m^3}{2n^2}}{\dfrac{3m}{8n^4}}$

14. _____

15. _____

16. $\dfrac{\dfrac{y^2-9}{2y^2-y-15}}{\dfrac{3y^2+10y+3}{2y^2+y-10}}$

17. $\dfrac{\dfrac{x^2+2xy+y^2}{x^2+3xy+2y^2}}{\dfrac{x^2-y^2}{x+2y}}$

16. _____

17. _____

In Problems 18 and 19, determine the domain of each rational function.

18. $R(x)=\dfrac{3x+2}{(4x-1)(x+5)}$

19. $R(x)=\dfrac{4x}{4x^2+1}$

18. _____

19. _____

20. If $f(x)=\dfrac{x^2-7x-8}{2x-5}$, $g(x)=\dfrac{2x^2+3x-20}{x^2-10x+16}$, and $h(x)=\dfrac{x^2-3x-40}{x+9}$, find

(a) $R(x)=f(x)\cdot g(x)$

(b) $R(x)=\dfrac{f(x)}{h(x)}$

20a. _____

20b. _____

Sullivan/Struve, *Intermediate Algebra*, 3e
Copyright © 2014 Pearson Education, Inc.

Five-Minute Warm-Up 5.2
Adding and Subtracting Rational Expressions

1. Determine the additive inverse of -21.

1. _____

2. Determine the least common denominator: $\dfrac{7}{24}$ and $\dfrac{14}{75}$.

2. _____

3. Using the rational numbers in problem 3, rewrite each number with LCD.

3. _____

In Problems 4 – 7, perform the indicated operation. Be sure to express the result in lowest terms.

4. $\dfrac{9}{4} + \dfrac{15}{4}$

5. $\dfrac{17}{30} + \dfrac{19}{45}$

4. _____

5. _____

6. $\dfrac{9}{28} - \dfrac{15}{28}$

7. $\dfrac{5}{16} - \left(-\dfrac{5}{24}\right)$

6. _____

7. _____

Guided Practice 5.2
Adding and Subtracting Rational Expressions

Objective 1: Add or Subtract Rational Expressions with a Common Denominator

1. To add or subtract rational expressions with a common denominator we use the following properties:

$$\frac{a}{c} + \frac{b}{c} = \frac{a+b}{c} \text{ and } \frac{a}{c} - \frac{b}{c} = \frac{a-b}{c}$$

where $\frac{a}{c}$ and $\frac{b}{c}$ are two rational expressions, $c \neq 0$, and then simplify the result.

Perform the indicated operation: $\frac{x^2 - 3}{2x + 3} + \frac{x^2 + x}{2x + 3}$, $x \neq -\frac{3}{2}$ *(See textbook Example 1)*

Step 1: Add the numerators and write the result over the common denominator

$$\frac{x^2 - 3}{2x + 3} + \frac{x^2 + x}{2x + 3} =$$

(a) _____

Combine like terms in the numerator:

(b) _____

Step 2: Simplify the rational expression.

Factor the numerator: **(c)** _____

Divide out common factors and express the answer as a simplified rational expression in factored form.

(d) _____

2. Perform the indicated operation: $\frac{x^2 - 11}{x^2 - 25} + \frac{-3x - 1}{25 - x^2}$, $x \neq 5$, $x \neq -5$ *(See textbook Examples 1 and 2)*

Step 1: Subtract the numerators and write the result over the common denominator

$$\frac{x^2 - 11}{x^2 - 25} + \frac{-3x - 1}{25 - x^2}$$

Factor -1 from $25 - x^2$ and use

$\frac{a}{-b} = \frac{-a}{b}$ to find the LCD.

Write the sum on the LCD.   **(a)** _____

Distribute the -1:   **(b)** _____

Combine like terms in the numerator: **(c)** _____

Step 2: Simplify the rational expression.

Factor the numerator and the denominator: **(d)** _____

Divide out common factors and express the answer as a simplified rational expression   **(e)** _____
in factored form.

Objective 2: Find the Least Common Denominator of Two or More Rational Expressions

3. List the steps for finding the least common denominator (LCD) of two or more rational expressions.

Objective 3: Add or Subtract Rational Expressions with Different Denominators

4. List the steps for adding or subtracting rational expressions with unlike denominators.

5. Add $\dfrac{3}{x+2} + \dfrac{8-2x}{x^2-4}$. Simplify the result, if possible. *(See textbook Example 5)*

(a) What is the LCD?   5a. _____

(b) Express each rational expression as an equivalent expression written on the common denominator.   5b. _____

(c) Add and simplify the result, if possible.

5c. _____

6. Subtract $\dfrac{a}{a^2+12a+20} - \dfrac{1}{a^2+8a-20}$. Simplify the result, if possible. *(See textbook Example 6)*

(a) What is the LCD?   6a. _____

(b) Express each rational expression as an equivalent expression written on the common denominator.   6b. _____

(c) Subtract and simplify the result, if possible.

6c. _____

7. Perform the indicated operations and simplify the result, if possible. *(See textbook Example 7)*

$$\frac{x}{x^2-2x+1} - \frac{2}{x} - \frac{x+1}{x^2-x^3}$$

(a) What is the LCD?   7a. _____

(b) Express each rational expression as an equivalent expression written on the common denominator.   7b. _____

(c) Perform the indicated operations and simplify the result, if possible.

7c. _____

Do the Math Exercises 5.2

Adding and Subtracting Rational Expressions

In Problems 1 – 4, perform the indicated operation and simplify the result.

1. $\dfrac{5x}{x-3} + \dfrac{2}{x-3}$

2. $\dfrac{9x}{6x-5} - \dfrac{2}{6x-5}$

1. _____

2. _____

3. $\dfrac{3x}{x-6} + \dfrac{2}{6-x}$

4. $\dfrac{x^2+2x-5}{x-4} - \dfrac{x^2-5x-15}{4-x}$

3. _____

4. _____

In Problems 5 – 7, find the least common denominator.

5. $\dfrac{1}{8a^3b}$ and $\dfrac{5}{12ab^2}$

6. $\dfrac{2m-7}{m^2+3m-18}$ and $\dfrac{5m+1}{m^2-7m+12}$

5. _____

6. _____

7. $\dfrac{x-6}{x^2-9}$ and $\dfrac{3x}{x^3-3x^2}$

7. _____

In Problems 8 – 17, add or subtract, as indicated, and simplify the result.

8. $\dfrac{2}{9x} + \dfrac{5}{3x^2}$

9. $\dfrac{x+2}{x-3} - \dfrac{x+2}{x+1}$

8. _____

9. _____

10. $\dfrac{z+1}{z+3} - \dfrac{z+17}{z^2-z-12}$

11. $\dfrac{x-5}{x^2+4x+3} + \dfrac{x-2}{x^2-1}$

10. _____

11. _____

12. $\dfrac{m-2n}{m^2+4mn+4n^2} + \dfrac{m-n}{m^2-mn-6n^2}$

13. $\dfrac{3}{x^2+7x+10} - \dfrac{4}{x^2+6x+5}$

12. _____

13. _____

14. $\dfrac{y^2+4y+4}{y^2-9} + \dfrac{y^2+4y+4}{9-y^2}$

15. $\dfrac{7}{m-3} - \dfrac{5}{m} - \dfrac{2m+6}{m^2-9}$

14. _____

15. _____

16. $\dfrac{2}{x} - \dfrac{2}{x+2} + \dfrac{2}{(x+2)^2}$

17. $3 + \dfrac{x+4}{x-4}$

16. _____

17. _____

18. Given that $f(x) = \dfrac{5}{x+2}$ and $g(x) = \dfrac{3}{x-1}$,

    (a) find $R(x) = f(x) + g(x)$

    (b) state the domain of $R(x)$

18a. _____

18b. _____

19. Surface Area of a Can The volume of a cylindrical can is 200 cubic centimeters. Its surface area S as a function of the radius r of the can is given by the function:

$$S(r) = 2\pi r^2 + \frac{400}{r}$$

    (a) Write S over a common denominator. That is, write S so that the rule is a single rational expression.

    (b) Find and interpret $S(4)$. Round your answer to two decimal places.

19a. _____

19b. _____

Sullivan/Struve, *Intermediate Algebra,* 3e
Copyright © 2014 Pearson Education, Inc.

Five-Minute Warm-Up 5.3
Complex Rational Expressions

1. _____

1. Factor $8x^2 + 2x - 3$.

In Problem 2 – 5, simplify each expression.

2. _____

2. $\dfrac{24p^4 q^2}{30pq^2}$ **3.** $\dfrac{36x^9 y^3}{80x^{12} y^2}$

3. _____

4. _____

4. $\left(-2r^3\right)^4$ **5.** $\dfrac{y^4 z^{-2}}{\left(y^4 z\right)^3}$

5. _____

In Problems 6 and 7, perform the indicated operation. Be sure to express the result in lowest terms.

6. _____

6. $\dfrac{\frac{4}{27}}{\frac{18}{54}}$ **7.** $\dfrac{18}{42} \div \left(-\dfrac{24}{49}\right)$

7. _____

Guided Practice 5.3
Complex Rational Expressions

Objective 1: Simplify a Complex Rational Expression by Simplifying the Numerator and Denominator Separately (Method I)

1. In your own words, define a *complex rational expression*.

2. Simplify $\dfrac{\dfrac{x}{x+1}}{1+\dfrac{1}{x-1}}$, $x \neq 1$, $x \neq -1$ using Method I. *(See textbook Examples 1 and 2)*

Step 1: Write the denominator of the complex rational expression as a single rational expression.

Determine the LCD of 1 and $x-1$:

(a) _____

Write the equivalent rational expressions on the LCD and then add using $\dfrac{a}{c}+\dfrac{b}{c}=\dfrac{a+b}{c}$:

(b) _____

Step 2: Write the numerator of the complex rational expression as a single rational expression.

This is already done.

Step 3: Rewrite the complex rational expression using the rational expressions determined in Steps 1 and 2.

(c) _____

Step 4: Simplify the rational expression using the techniques for dividing rational expressions from Section 5.1

Rewrite the division problem as a multiplication problem:

(d) _____

Divide out common factors and express the answer as a simplified rational expression in factored form.

(e) _____

Objective 2: Simplify a Complex Rational Expression Using the Least Common Denominator (Method II)

3. Method II uses the LCD to simplify complex rational expressions. We use several of the properties of real numbers to simplify the complex rational expression; that is, to find an equivalent rational expression which has a single fraction bar. State the property of real numbers that is illustrated below.

(a) $\dfrac{x-9}{x-9}=1$

3a. _____

(b) $\dfrac{2}{x} \cdot \dfrac{x-9}{x-9}=\dfrac{2}{x}$

3b. _____

(c) $\dfrac{2}{x} \cdot \dfrac{x-9}{x-9}=\dfrac{2x-18}{x^2-9x}$

3c. _____

4. Simplify $\dfrac{\dfrac{x^2}{x^2-16}-\dfrac{x}{x+4}}{\dfrac{x}{x^2-16}-\dfrac{1}{x-4}}$, $x \neq 4$, $x \neq -4$ using Method II. *(See textbook Examples 3 and 4)*

Step 1: Find the least common denominator among all the denominators in the complex rational expression.

Determine the LCD of x^2-16, $x+4$ and $x-4$:

(a) _____

Step 2: Multiply both the numerator and denominator of the complex rational expression by the LCD found in Step 1.

(b) _____

Distribute the LCD to each term:

(c) _____

Step 3: Simplify the rational expression.

Divide out the common factors:

(d) _____

Simplify the expression:

(e) _____

5. Simplify $\dfrac{x^{-1}+3^{-1}}{x^{-2}-9^{-1}}$ as a rational expression that contains no negative exponents. *(See textbook Example 5)*

(a) Rewrite the expression as a complex rational expression which does not contain negative exponents.

5a. _____

(b) Simplify by either Method I or Method II.

5b. _____

Do the Math Exercises 5.3
Complex Rational Expressions

In Problems 1 – 14, simplify the complex rational expression using either Method I or Method II.

1. $\dfrac{1 + \dfrac{1}{x^2}}{1 - \dfrac{1}{x^2}}$

2. $\dfrac{\dfrac{7}{w} + \dfrac{9}{x}}{\dfrac{9}{w} - \dfrac{7}{x}}$

1. _____

2. _____

3. $\dfrac{\dfrac{a}{a+1} - 1}{\dfrac{a+3}{a} - 2}$

4. $\dfrac{\dfrac{x+5}{x-2} - \dfrac{x+3}{x-1}}{3x+1}$

3. _____

4. _____

5. $\dfrac{\dfrac{x-4}{x-1} - \dfrac{x}{x-3}}{3 + \dfrac{12}{x-3}}$

6. $\dfrac{1 - \dfrac{4}{z}}{z - \dfrac{16}{z}}$

5. _____

6. _____

7. $\dfrac{\dfrac{n^2}{m} - \dfrac{m^2}{n}}{\dfrac{1}{m} - \dfrac{1}{n}}$

8. $\dfrac{1 + \dfrac{5}{x}}{1 + \dfrac{1}{x+4}}$

7. _____

8. _____

9. $\dfrac{\dfrac{5}{x} - \dfrac{x}{5}}{\dfrac{1}{5} - \dfrac{5}{x^2}}$

10. $\dfrac{\dfrac{x-3}{x+3} + \dfrac{x-3}{x-4}}{1 + \dfrac{x+3}{x-4}}$

9. _____

10. _____

11. $\dfrac{\dfrac{-6}{x^2 + 5x + 6}}{\dfrac{2}{x+3} - \dfrac{3}{x+2}}$

12. $\dfrac{2x^{-1} + 2y^{-1}}{xy^{-1} - x^{-1}y}$

11. _____

12. _____

13. $\dfrac{(x-y)^{-1}}{x^{-1} - y^{-1}}$

14. $\dfrac{a^{-3} + 8b^{-3}}{a^{-2} - 4b^{-2}}$

13. _____

14. _____

15. Electric Circuits An electric circuit contains three resistors connected in parallel. If the resistance of each is R_1, R_2, and R_3 ohms, respectively, then their combined resistance is given by the formula

$$R = \dfrac{1}{\dfrac{1}{R_1} + \dfrac{1}{R_2} + \dfrac{1}{R_3}}$$

(a) Express R as a simplified rational expression.

(b) Evaluate the rational expression if $R_1 = 4$ ohms, $R_2 = 6$ ohms and $R_3 = 10$ ohms.

15a. _____

15b _____

Five-Minute Warm-Up 5.4
Rational Equations

1. Solve: $\dfrac{3x + 5}{2} = 4 + \dfrac{4x + 2}{4}$

    1. _____

2. Solve: $-3p^2 + 6p = -72$

    2. _____

3. Factor: $3z^2 - 14z + 8$

    3. _____

4. Determine which of the following are in the domain of the variable x for the expression

    4. _____

$$\frac{x + 4}{x^2 + 2x - 8}.$$

    (a) $x = -4$ **(b)** $x = -2$ **(c)** $x = 2$

5. Given $f(x) = x^2 - 12x - 45$ solve $f(x) = 0$.

    5. _____

6. If $g(-2) = 12$, what is the point on the graph of g?

    6. _____

Guided Practice 5.4
Rational Equations

Objective 1: Solve Equations Containing Rational Expressions

1. When solving rational equations it is important to identify the domain of the variable. We exclude all

values of the variable that result in _____.

2. Solve: $\dfrac{x+5}{x-7} = \dfrac{x-3}{x+7}$ *(See textbook Example 1)*

Step 1: Determine the domain of the variable in the rational equation.

For what values of x will either denominator be equal to zero? **(a)** _____

Express the domain of x: **(b)** _____

Step 2: Determine the least common denominator (LCD) of all the denominators.

LCD: **(c)** _____

Step 3: Multiply both sides of the equation by the LCD and simplify the expression on each side of the equation.

Multiply both sides by the LCD: **(d)** _____

Divide out like factors: **(e)** _____

Multiply: **(f)** _____

Step 4: Solve the resulting equation.

Solve for x: **(g)** _____

Step 5: Check Verify your solution using the original equation.

We leave it to you to verify your solution.

Write the solution set: **(h)** _____

3. Solve: $\dfrac{x}{12} + \dfrac{1}{2} = \dfrac{1}{3x} + \dfrac{2}{x^2}$ *(See textbook Example 2)*

(a) Determine the domain of the variable. 3a. _____

(b) Determine the LCD of all of the denominators. 3b. _____

(c) Multiply both sides of the equation by the LCD. What is the resulting equation?

3c. _____

(d) Solve the resulting equation. 3d. _____

4. Solve: $\dfrac{6}{x^2 - 1} = \dfrac{5}{x - 1} - \dfrac{3}{x + 1}$ *(See textbook Example 4)*

(a) Determine the domain of the variable. 4a. _____

(b) Determine the LCD of all of the denominators. 4b. _____

(c) Multiply both sides of the equation by the LCD and solve the resulting equation. 4c. _____
What is the solution to this equation?

(d) What is the solution set? 4d. _____

5. Solve: $\dfrac{5}{x - 4} + \dfrac{3}{x - 2} = \dfrac{x^2 - x - 2}{x^2 - 6x + 8}$ *(See textbook Example 5)*

(a) Determine the domain of the variable. 5a. _____

(b) Determine the LCD of all of the denominators. 5b. _____

(c) Multiply both sides of the equation by the LCD and solve the resulting equation. 5c. _____
What is the solution to this equation?

(d) What is the solution set? 5d. _____

Objective 2: Solve Equations Involving Rational Functions

6. For the function $f(x) = x + \dfrac{7}{x}$, $f(x) = 8$, what point(s) are on the graph of *f*? *(See textbook Example 6)*

(a) To begin, we write the equation: 6a. _____

(b) Determine the domain of the variable. 6b. _____

(c) Determine the LCD of all of the denominators. 6c. _____

(d) Multiply both sides of the equation by the LCD and solve the resulting equation. 6d. _____
What is the solution to this equation?

(e) What is the solution set? 6e. _____

(f) Write the ordered pair(s) that are on the graph of *f*. 6f. _____

Sullivan/Struve, *Intermediate Algebra, 3e*
Copyright © 2014 Pearson Education, Inc.

Do the Math Exercises 5.4
Rational Equations

In Problems 1 – 12, solve each equation. Be sure to verify your results.

1. $\dfrac{8}{p} + \dfrac{1}{4p} = \dfrac{11}{8}$

2. $\dfrac{2x+1}{x+3} = \dfrac{4(x-1)}{2x+3}$

1. _____

2. _____

3. $m + \dfrac{8}{m} = 6$

4. $8b - \dfrac{3}{b} = 2$

3. _____

4. _____

5. $2 - \dfrac{3}{p+2} = \dfrac{6}{p}$

6. $\dfrac{5}{x+2} = 1 - \dfrac{3}{x-2}$

5. _____

6. _____

7. $\dfrac{4}{x+3} + \dfrac{5}{x-6} = \dfrac{4x+1}{x^2-3x-18}$

8. $\dfrac{3}{x-4} = \dfrac{5x+4}{x^2-16} - \dfrac{4}{x+4}$

7. _____

8. _____

9. $p + \dfrac{25}{p} = 10$

10. $\dfrac{3}{a^2+3a-10} + \dfrac{2}{a^2+7a+10} = \dfrac{4}{a^2-4}$

9. _____

10. _____

11. $\dfrac{9}{b} + \dfrac{4}{5b} = \dfrac{7}{10}$

12. $\dfrac{x+3}{x-2} + 4 = \dfrac{x+2}{x+1}$

11. _____

12. _____

13. Solve: $2 + 11a^{-1} = -12a^{-2}$

13. _____

14. For the function $f(x) = 2x + \dfrac{8}{x}$, solve $f(x) = -10$. What point(s) are on the graph of f?

14. _____

_____

15. Let $f(x) = \dfrac{4x+1}{8x+5}$ and $g(x) = \dfrac{x-4}{2x-7}$. For what value(s) of x does $f(x) = g(x)$?

What are the point(s) of intersection of the graphs of f and g?

15. _____

_____

In Problems 16 – 19, simplify or solve.

16. $\left(\dfrac{z^{-2}}{2z^{-3}}\right)^{-1} + 3(z-1)^{-1}$

17. $\dfrac{5}{x-6} + \dfrac{2}{x+2} = \dfrac{1}{x^2 - 4x - 12}$

16. _____

17. _____

18. $\dfrac{2a^6}{\left(a^3\right)^2} - \dfrac{5a^2}{a^3} = \dfrac{3a}{a^3}$

19. $\dfrac{3}{x-2} - \dfrac{2x+1}{x+1}$

18. _____

19. _____

Five-Minute Warm-Up 5.5
Rational Inequalities

1. Write in interval notation: $-3 \leq x < 2$

1. _____

2. Solve and graph the solution set: $-5x + 1 > 9 + 3(2x + 1)$

2. _____

3. Solve and write the solution set in interval notation: $\dfrac{3z - 1}{4} - 1 \leq \dfrac{6z + 5}{2}$

3. _____

4. Determine whether $x = -2$ satisfies the inequality $5x + 17 \geq 7$.

4. _____

Guided Practice 5.5
Rational Inequalities

Objective 1: Solve a Rational Inequality

1. Solving rational inequalities depends on finding intervals where the polynomials in the rational expression are positive or negative. This means that the inequality must be of the form:

   (a) $\dfrac{p(x)}{q(x)} >$ _____ or $\dfrac{p(x)}{q(x)} \geq$ _____ when the quotient is positive and

   (b) $\dfrac{p(x)}{q(x)} <$ _____ or $\dfrac{p(x)}{q(x)} \leq$ _____ when the quotient is negative.

2. The quotient is positive when both $p(x)$ and $q(x)$ are _____ or when both $p(x)$ and $q(x)$ are _____ .

3. The quotient is negative when $p(x)$ is _____ and $q(x)$ is _____ or when $p(x)$ is _____ and $q(x)$ is _____ .

4. We always exclude the values that cause _____ .

5. Solve $\dfrac{x+2}{x-4} \leq 0$. Graph the solution set. *(See textbook Example 1)*

Step 1: Write the inequality so that a rational expression is on one side of the inequality and zero is on the other. Be sure to write the rational expression as a single quotient.

$\dfrac{x+2}{x-4} \leq 0$ is already in the required form.

Step 2: Determine the numbers for which the rational expression equals 0 or is undefined.

Value(s) when the rational expression = 0:

(a) _____

Value(s) when the rational expression is undefined:

(b) _____

Step 3: Use the numbers found in Step 2 to separate the real number line into intervals.

Step 4: Choose a test point within each interval formed in Step 3 to determine the sign of $x+2$ and $x-4$. Then determine the sign of the quotient.

Select a test point in the interval $(-\infty, -2)$:

(c) _____

Sign of $x + 2$:

(d) _____

Sign of $x - 4$:

(e) _____

Sign of the quotient:

(f) _____

continued next page

Select a test point in the
interval $(-2, 4)$:　　　　　　　　**(g)** _____

Sign of $x + 2$:　　　　　　　　**(h)** _____

Sign of $x - 4$:　　　　　　　　**(i)** _____

Sign of the quotient:　　　　　**(j)** _____

Select a test point in the
interval $(4, \infty)$:　　　　　　　**(k)** _____

Sign of $x + 2$:　　　　　　　　**(l)** _____

Sign of $x - 4$:　　　　　　　　**(m)** _____

Sign of the quotient:　　　　　**(n)** _____

In this problem, are we
looking for the quotient to be **(o)** _____
positive or negative?

What value must be excluded? **(p)** _____

Graph the solution set:　　　　**(q)**

6. Rewrite the rational expression $\dfrac{5}{x-1} - \dfrac{7}{x+1} > 0$ as single rational expression which
is positive.

6. _____

7. Rewrite the rational expression $\dfrac{3x-7}{x+2} < 2$ as single rational expression which
is negative.

7. _____

8. Complete the following chart used to solve $\dfrac{(x+4)(x-3)}{x-2} \geq 0$. *(See textbook Example 2)*

Interval	$(-\infty, -4)$	-4	$(-4, 2)$	2	$(2, 3)$	3	$(3, \infty)$
Test Point							
Sign of $x + 4$							
Sign of $x - 3$							
Sign of $x - 2$							
Sign of quotient							
Conclusion							

9. Graph the solution set to Problem **8**.

Sullivan/Struve, *Intermediate Algebra,* 3e
Copyright © 2014 Pearson Education, Inc.

Do the Math Exercises 5.5
Rational Inequalities

In Problems 1 – 10, solve each rational inequality. Express your answer in interval notation.

1. $\dfrac{x+8}{x+2} < 0$

2. $\dfrac{x+12}{x-2} \geq 0$

1. _____

2. _____

3. $\dfrac{x-10}{x+5} \leq 0$

4. $\dfrac{(5x-2)(x+4)}{x-5} < 0$

3. _____

4. _____

5. $\dfrac{(3x-2)(x-6)}{x+1} \geq 0$

6. $\dfrac{x+3}{x-4} > 1$

5. _____

6. _____

7. $\dfrac{3x-7}{x+2} \leq 2$

8. $\dfrac{3x+20}{x+6} < 5$

7. _____

8. _____

9. $\dfrac{2}{x+3} + \dfrac{2}{x} \leq 0$

10. $\dfrac{1}{x-4} \geq \dfrac{3}{2x+1}$

9. _____

10. _____

In Problems 11 and 12, for each function find the values of x that satisfy the given condition.

11. Solve $R(x) \geq 0$ if $R(x) = \dfrac{x+3}{x-8}$ **12.** Solve $R(x) < 0$ if $R(x) = \dfrac{3x+2}{x-4}$ 11. _____

12. _____

In Problems 13 and 14, find the x-intercept of the graph of each function.

13. $h(x) = -3x^2 - 7x + 20$ **14.** $R(x) = \dfrac{x^2 + 5x + 6}{x+2}$ 13. _____

14. _____

15. Average Cost Suppose that the daily cost C of manufacturing x bicycles is given by $C(x) = 90x + 5000$. Then the average daily cost $\overline{C}$ is given by $\overline{C}(x) = \dfrac{90x + 5000}{x}$. How many bicycles must be produced each day in order for the average cost to be no more than \$130?   15. _____

Sullivan/Struve, *Intermediate Algebra, 3e*
Copyright © 2014 Pearson Education, Inc.

Five-Minute Warm-Up 5.6
Models Involving Rational Expressions

1. Solve for x: $4x + 3y = -24$

1.

2. Solve for y: $\dfrac{9x + 2y}{3} = 12$

2.

3. **Train Station** Two trains leave a station at the same time. One train is traveling east at 10 mph faster than the other train, which is traveling west. After 6 hours, the two trains are 720 miles apart. At what speed did the faster train travel?

3. _____

Guided Practice 5.6
Models Involving Rational Expressions

Objective 1: Solve for a Variable in a Rational Expression

1. The expression "to solve for the variable" means to get the variable by itself on one side of the equation with all other variables and constants, if any, on the other side. Solve for the indicated variable. *(See textbook Example 1)*

$\dfrac{2}{x} - \dfrac{1}{y} = \dfrac{6}{z}$ for y

1. _____

Objective 2: Model and Solve Ratio and Proportion Problems

2. Write an example of a ratio.

2. _____

3. Write an example of a proportion.

3. _____

4. Solve: $\dfrac{5}{k+4} = \dfrac{2}{k-1}$

4. _____

5. In your own words, what does it mean if two geometric figures are *similar*?

6. Suppose that a 6-foot tall man casts a show of 3.2 feet. At the same time of day, a tree casts a shadow of 8 feet. How tall is the tree? *(See textbook Example 2)*

(a) Write a proportion that can be used to solve this problem.

6a. _____

(b) Solve the proportion. How tall is the tree?

6b. _____

Objective 3: Model and Solve Work Problems

7. Problems of this type involve completing a job or a task when working at a constant rate. We convert the time t that it takes to complete the job into a unit rate. That is, if it takes t hours to complete a job, then $\dfrac{1}{t}$ of the job is completed per hour.

(a) If it takes 12 minutes to complete the job, what part of the job is completed per minute?   7a. _____

(b) If it takes x hours to complete the job, what part of the job is completed per hour?   7b. _____

(c) If it takes $t+2$ hours to complete the job, what part of the job is competed per hour?   7c. _____

8. Josh can clean the math building on his campus in 3 hours. Ken takes 5 hours to clean the same building. If they work together, how long will it take for Josh and Ken to clean the math building? *(See textbook Example 4)*

Step 1: Identify We want to know how long it will take Josh and Ken working together to clean the building.

Step 2: Name We let t represent the time (in hours) that it takes to clean the building when working together.

Step 3: Translate What fraction of the job is completed in one hour when working individually and when working together? Write the following ratios:

(a) Part of the job completed by Josh in one hour: 8a. _____

(b) Part of the job completed by Ken in one hour: 8b. _____

(c) Part of the job completed when working together in one hour: 8c. _____

(d) Write the model for this problem:
                                                                          8d. _____
Step 4: Solve the equation from Step 3.

Step 5: Check Is your answer reasonable?

(e) Step 6: Answer the question.   8e. _____

Objective 4: Model and Solve Uniform Motion Problems

9. A small plane can travel 1000 miles with the wind in the same time it can go 600 miles against the wind. If the speed of the plane in still air is 180 mph, what is the speed of the wind? *(See textbook Example 6)*

(a) Rate (in mph) when traveling with the wind: 9a. _____

(b) Rate (in mph) when traveling against the wind: 9b. _____

(c) Use $t = \dfrac{d}{r}$ to write a rational expression for the time traveled with the wind:
                                                                          9c. _____

(d) Use $t = \dfrac{d}{r}$ to write a rational expression for the time traveled against the wind:
                                                                          9d. _____

(e) Write an equation that can be used to solve this problem:
                                                                          9e. _____

(f) Solve your equation and answer the question.   9f. _____

Sullivan/Struve, *Intermediate Algebra*, 3e
Copyright © 2014 Pearson Education, Inc.

Do the Math Exercises 5.6
Models Involving Rational Expressions

In Problems 1 – 4, solve each formula for the indicated variable.

1. Solve $\dfrac{V_1}{V_2} = \dfrac{P_2}{P_1}$ for V_2.

2. Solve $P = \dfrac{A}{1+r}$ for r.

1. _____

2. _____

3. Solve $m = \dfrac{y - y_1}{x - x_1}$ for x_1.

4. Solve $v_2 = \dfrac{2m_1 v_1}{m_1 + m_2}$ for m_1.

3. _____

4. _____

5. **Flight Accidents** According to the Statistical Abstract of the United States, in 2010, there were 1.27 fatal airplane accidents per 100,000 flight hours. Also, in 2010, there were a total of 267 fatal accidents. How many flight hours were flown in 2010?

5.

6. **Painting a Room** Latoya can paint five 10 foot by 14 foot rooms by herself in 14 hours. Lisa can paint five 10 foot by 14 foot rooms by herself in 10 hours. Working together, how long would it take to paint five 10 foot by 14 foot rooms?

6. _____

7. **Assembling a Swing Set** Alexandra and Frank can assemble a King Kong swing set working together in 6 hours. One day, when Frank called in sick, Alexandra was able to assemble a King Kong swing set in 10 hours. How long would it take Frank to assemble a King Kong swing set if he worked by himself?

7.

8. **Running a Race** Roger can run one mile in 8 minutes. Jeff can run one mile in 6   8. _____
minutes. If Jeff gives Roger a 1 minute head start, how long will it take before Jeff
catches up to Roger? How far will each have run?

9. **Draining a Pond** A pond can be emptied in 3.75 $\left(= \dfrac{15}{4} \right)$ hours using a 10 horsepower   9. _____

pump along with a 4 horsepower pump. The 4 horsepower pump requires 4 hours more
than the 10 horsepower pump to empty the pond when working by itself. How long
would it take to empty the pond using just the 10 horsepower pump?

10. **Round trip** A plane flies 600 miles west (into the wind) and makes the return trip   10. _____
following the same flight path. The effect of the jet stream on the plane is 15 miles per
hour. The roundtrip takes 9 hours. What is the speed of the plane in still air?

11. **Riding Your Bicycle** Every weekend, you ride your bicycle on a forest preserve path.   11. _____
The path is 20 miles long and ends at a waterfall, at which point you relax and then make
the trip back to the starting point. One weekend, you find that in the same time it takes
you to travel to the waterfall you are only able to return 12 miles. Your average speed
going to the waterfall is 4 miles per hour faster than your average speed on the return trip.
What was your average speed going to the waterfall?

Sullivan/Struve, *Intermediate Algebra*, 3e
Copyright © 2014 Pearson Education, Inc.

Name:

Instructor:

Date:

Section:

Five-Minute Warm-Up 5.7
Variation

1. Solve each equation.

   (a) $-75 = 15x$

   (b) $-\dfrac{2}{3}k = \dfrac{8}{27}$

   1a. _____

   1b. _____

2. Solve: $-12 = \dfrac{k}{4}$

   2. _____

3. Graph the equation $y = \dfrac{1}{2}x$.

4. Graph the equation $y = 2x$.

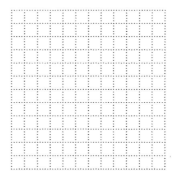

Guided Practice 5.7
Variation

Objective 1: Model and Solve Direct Variation Problems

1. We say that y varies directly with x, or y is directly proportional to x, if there is a nonzero number k such

that _____.

2. The number k is called the _____.

3. If y varies directly with x, then y is a _____ function of x and has a y- intercept of _____.

4. Suppose that y is directly proportional to x and when $x = -12$, $y = 5$. Find y when $x = 20$. Write the direct variation equation, calculate the constant k, and then use the given values to solve the unknown.

4. _____

Objective 2: Model and Solve Inverse Variation Problems

5. We say that y varies inversely with x, or y is inversely proportional to x, if there is a nonzero number k

such that _____.

6. Suppose that y varies inversely with x. When $x = 4$, $y = 12$. Find y when $x = 18$. Write   6. _____
the inverse variation equation, calculate the constant, k, and then use the given values to
solve for the unknown. *(See textbook Example 3)*

Objective 3: Model and Solve Joint Variation and Combined Variation Problems

7. Suppose that r varies jointly with s and t. When $r = 12$, $s = 8$ and $t = 3$. Find r when   7. _____
$s = 14$ and $t = 6$. Write the joint variation equation, calculate the constant, k, and then use the given values to solve for the unknown. *(See textbook Example 4)*

8. Gas Laws The volume V of an ideal gas varies directly with the temperature T and inversely with the pressure P. If a cylinder contains oxygen at a temperature of 300 kelvin (K) and a pressure of 15 atmospheres (Atm) in a volume of 100 liters, what is the constant of proportionality k? If a piston is lowered into the cylinder, decreasing the volume occupied by the gas to 70 liters and raising the temperature to 315 K, what is the pressure?

(a) Write the equation that shows the relationship between the variables, V, T, and P.   8a. _____

(b) To calculate the constant of proportionality, substitute each of the following variables   8b.
into your equation from 8a. $T =$ _____

      $P =$ _____

      $V =$ _____

(c) Solve your equation to determine the constant of proportionality, k.   8c. _____

(d) Substitute your value of k into 8a to determine the function relating the variables.   8d. _____

(e) Substitute the values of V and T in the last sentence into your equation from 8d to answer
the question.   8e. _____

      Sullivan/Struve, *Intermediate Algebra*, 3e
Copyright © 2014 Pearson Education, Inc.

Do the Math Exercises 5.7
Variation

In Problems 1 and 2, (a) find the constant of proportionality, k, (b) write the linear function relating the two variables, and (c) find the quantity indicated.

1. Suppose that *y* varies directly with *x*. When $x = 3$, then $y = 15$. Find *y* when $x = 5$.

2. Suppose that *y* is directly proportional to *x*. When $x = 20$, then $y = 4$. Find *y* when $x = 35$.

1a. _____

1b. _____

1c. _____

2a. _____

2b. _____

2c. _____

In Problems 3 and 4, find the quantity indicated.

3. *A* is directly proportional to *B*. If *A* is 360 when *B* is 72, find *B* when *A* is 400.

4. *m* varies directly with *r*. If *r* is 24 when *m* is 9, find *r* when *m* is 24.

3. _____

4. _____

5. **Mortgage Payments** The monthly payment *p* on a mortgage varies directly with the amount borrowed *b*. Suppose that you decide to borrow $120,000 using a 15-year mortgage at 5.5% interest. You are told that your payment is $980.50. Assume that you have decided to buy a more expensive home that requires you borrow $150,000. What will your monthly payment be?

5. _____

6. **Buying Gasoline** The cost to purchase a tank of gasoline varies directly with the number of gallons purchased. You notice that the person in front of you spent $34.50 on 15 gallons of gas. If your SUV needs 35 gallons of gas, how much will you spend?

6. _____

7. **Falling Objects** The velocity of a falling object (ignoring air resistance) v is directly proportional to the time t of the fall. If, after 2 seconds, the velocity of the object is 64 feet per second, what will its velocity be after 3 seconds?

7. _____

*In Problems 8 – 10, **(a)** find the constant, k; **(b)** write the function relating the two variables; and **(c)** find the quantity indicated.*

8. Suppose that y is inversely proportional to x. When $x = 20$, then $y = 4$. Find y when $x = 35$.

9. Suppose that y varies jointly with x and z. When $y = 20$, $x = 6$ and $z = 10$. Find y when $x = 8$ and $z = 15$.

8a. _____

8b. _____

8c. _____

9a. _____

9b. _____

9c. _____

10. Suppose that Q varies directly with x and inversely with y. When $Q = \dfrac{14}{5}$, $x = 4$ and $y = 3$. Find Q when $x = 8$ and $y = 3$.

10a. _____

10b. _____

10c. _____

11. **Resistance** The current, i, in a circuit is inversely proportional to its resistance, R, measured in ohms. Suppose that when the current in a circuit is 30 amperes, the resistance is 8 ohms. Find the current in the same circuit when the resistance is 10 ohms.

11. _____

12. **Intensity of Light** The intensity, I, of light (measured in foot-candles) varies inversely with the square of the distance from the bulb. Suppose the intensity of a 100-watt light bulb at a distance of 2 meters is 0.075 foot-candles. Determine the intensity of the bulb at a distance of 3 meters.

12. _____

Five-Minute Warm-Up 6.1
*n*th Roots and Rational Exponents

In Problems 1 – 4, simplify each expression.

1. $\sqrt{0.25}$

2. $\left(\sqrt{\dfrac{4}{9}}\right)^2$

1. _____

2. _____

3. $\sqrt{(4x+3)^2}$

4. $\sqrt{x^2 - 2xy + y^2}$

3. _____

4. _____

In Problems 5 – 8, simplify each expression.

5. 7^{-2}

6. $\dfrac{1}{x^{-2}}$

5. _____

6. _____

7. $\left(\dfrac{3x}{y^2}\right)^2$

8. $\left(\dfrac{4a^2b^{-1}}{8a^2b^3}\right)^{-3}$

7. _____

8. _____

Guided Practice 6.1
*n*th Roots and Rational Exponents

Objective 1: Evaluate nth Roots

1. In the notation $\sqrt[3]{8} = 2$, 3 is called the _____, 8 is called the _____ and 2 is the _____.

2. If the index is even, then the radicand must be _____ in order for the radical to simplify to a real number.

If the index is odd, then the radicand can be _____ and the expression will simplify to any real number.

3. Since $(-4)^2 = 16$ and $4^2 = 16$, it could be interpreted that $\sqrt{16} = -4$ or 4. In fact, $\sqrt{16} = 4$ only, because when the index even, we use the _____ _____, which must be ≥ 0.

4. Evaluate each root without using a calculator. *(See textbook Example 1)*

(a) $\sqrt[3]{-27}$ (b) $\sqrt{-36}$ (c) $\sqrt[4]{\dfrac{16}{81}}$

4a. _____

4b. _____

4c. _____

5. Write $\sqrt[4]{64}$ as a decimal rounded to two decimal places. *(See textbook Example 2)*   5. _____

Objective 2: Simplify Expressions of the form $\sqrt[n]{a^n}$

6. Simplify: $\sqrt[4]{(2x-1)^4}$ *(See textbook Example 3)*   6. _____

Objective 3: Evaluate Expressions of the Form $a^{\frac{1}{n}}$

7. Write each of the following expressions as a radical and simplify, if possible. *(See textbook Example 4)*

(a) $144^{\frac{1}{2}}$ (b) $(-64)^{\frac{1}{2}}$ (c) $2x^{\frac{1}{3}}$

7a. _____

7b. _____

7c. _____

Objective 4: Evaluate Expressions of the Form $a^{\frac{m}{n}}$

8. If a is a real number and $\dfrac{m}{n}$ is a rational number in lowest terms with $n \geq 2$, then

$\qquad a^{\frac{m}{n}} =$ _____ or _____, provided that $\sqrt[n]{a}$ exists.

9. Evaluate each of the following expressions, if possible. *(See textbook Example 6)*

(a) $36^{\frac{3}{2}}$ **(b)** $-16^{\frac{5}{2}}$ **(c)** $(-64)^{\frac{2}{3}}$

9a. _____

9b. _____

9c. _____

10. Write $25^{\frac{2}{3}}$ as a decimal rounded to two decimal places. *(See textbook Example 7)*

10. _____

11. Write each radical expression with a rational exponent. *(See textbook Example 8)*

(a) $\sqrt[4]{(2x)^3}$ **(b)** $\left(\sqrt[3]{2x^2 y}\right)^2$

11a. _____

11b. _____

12. Rewrite each of the following with positive exponents, and completely simplify, if possible. *(See textbook Example 9)*

(a) $64^{-\frac{1}{2}}$ **(b)** $\dfrac{2}{16^{-\frac{1}{2}}}$ **(c)** $(4x)^{-\frac{5}{2}}$

12a. _____

12b. _____

12c. _____

Sullivan/Struve, *Intermediate Algebra*, 3e

Do the Math Exercises 6.1
nth Roots and Rational Exponents

In Problems 1 – 7, simplify each radical.

1. $\sqrt[3]{216}$

2. $\sqrt[3]{-64}$

3. $-\sqrt[4]{256}$

4. $\sqrt[3]{\dfrac{8}{125}}$

5. $\sqrt[4]{6^4}$

6. $\sqrt[5]{n^5}$

7. $\sqrt[6]{(2x-3)^6}$

1. _____

2. _____

3. _____

4. _____

5. _____

6. _____

7. _____

In Problems 8 – 16, evaluate each expression, if possible.

8. $16^{\frac{1}{2}}$

9. $-25^{\frac{1}{2}}$

10. $-81^{\frac{1}{4}}$

11. $(-81)^{\frac{1}{2}}$

12. $-100^{\frac{5}{2}}$

13. $-(-32)^{\frac{3}{5}}$

14. $121^{-\frac{1}{2}}$

15. $\dfrac{1}{49^{-\frac{3}{2}}}$

16. $27^{-\frac{4}{3}}$

8. _____

9. _____

10. _____

11. _____

12. _____

13. _____

14. _____

15. _____

16. _____

In Problems 17 – 19, rewrite each of the following radicals with a rational exponent.

17. $\sqrt[4]{x^3}$

18. $\left(\sqrt[5]{3x}\right)^2$

17. _____

18. _____

19. $\sqrt[4]{(3pq)^7}$

19. _____

In Problems 20 – 21, use a calculator to write each expression as a decimal rounded to two decimal places.

20. $\sqrt[3]{85}$

21. $100^{0.25}$

20. _____

21. _____

In Problems 22 – 26, evaluate each expression, if possible.

22. $100^{\frac{3}{2}}$

23. $\sqrt[4]{-1}$

22. _____

23. _____

24. $125^{-\frac{1}{3}}$

25. $100^{\frac{1}{2}} - 4^{\frac{3}{2}}$

24. _____

25. _____

26. $(-125)^{-\frac{1}{3}}$

26. _____

27. Explain why $(-9)^{\frac{1}{2}}$ is not a real number, but $-9^{\frac{1}{2}}$ is a real number.

Sullivan/Struve, *Intermediate Algebra*, 3e
Copyright © 2014 Pearson Education, Inc.

Five-Minute Warm-Up 6.2
Simplifying Expressions Using the Laws of Exponents

In Problems 1 – 8, simplify each expression.

1. $9z^6 \bullet 6z$

2. $\dfrac{18u^5}{12u^3}$

1. _____

2. _____

3. 25^{-2}

4. $\left(\dfrac{4}{3}\right)^{-3}$

3. _____

4. _____

5. $\dfrac{9}{7}x^5 y^{-2} \bullet \dfrac{28}{12}xy$

6. $\left(-5p^3\right)^{-2}$

5. _____

6. _____

7. $\left(\dfrac{8a^{-1}}{b^{-5}}\right)^2$

8. $\left(\dfrac{2x^{-2}y}{6x^{-3}y^2}\right)^{-1} \cdot \left(\dfrac{xy^{-2}}{9xy^2}\right)^2$

7. _____

8. _____

9. Evaluate: $\sqrt{\dfrac{81}{49}}$

9. _____

Guided Practice 6.2
Simplifying Expressions Using the Laws of Exponents

Objective 1: Simplify Expressions Involving Rational Exponents

1. List four characteristics of an expression with rational exponents which is completely simplified.

2. Simplify each of the following expressions involving rational exponents. Express the answer with positive rational exponents, if necessary. *(See textbook Examples 1 and 2)*

(a) $6^{\frac{12}{5}} \cdot 6^{-\frac{2}{5}}$ **(b)** $\dfrac{\left(-8a^4\right)^{\frac{1}{3}}}{a^{\frac{5}{6}}}$

2a. _____

2b. _____

3. Simplify the following expression involving rational exponents. Express the answer with positive rational exponents, if necessary. *(See textbook Example 3)*

$\left(xy^{-\frac{3}{8}}\right) \cdot \left(x^{\frac{1}{2}}y^{-2}\right)^{\frac{3}{4}}$

3. _____

Objective 2: Simplify Radical Expressions

4. Rewrite the radical as an expression involving a rational exponent and simplify. Express the answer with a simplified radical, if necessary. *(See textbook Example 4)*

(a) $\sqrt[6]{81^3}$

(b) $\sqrt[4]{128x^8y^4}$

(c) $\dfrac{\sqrt[4]{x^3}}{\sqrt{x}}$

(d) $\sqrt[3]{\sqrt{p}}$

4a. _____

4b. _____

4c. _____

4d. _____

Objective 3: Factor Expressions Containing Rational Exponents

5. Simplify $6x^{\frac{2}{3}} + 2x^{\frac{1}{3}}(5x - 2)$ by factoring out $2x^{\frac{1}{3}}$. Express the answer with positive rational exponents, if necessary. *(See textbook Example 5)*

5. _____

Sullivan/Struve, *Intermediate Algebra, 3e*
Copyright © 2014 Pearson Education, Inc.

Do the Math Exercises 6.2
Simplifying Expressions Using the Laws of Exponents

In Problems 1 – 9, simplify each of the following expressions.

1. $3^{\frac{1}{3}} \cdot 3^{\frac{5}{3}}$

2. $\dfrac{10^{\frac{7}{5}}}{10^{\frac{2}{5}}}$

1. _____

2. _____

3. $\dfrac{y^{\frac{1}{5}}}{y^{\frac{9}{10}}}$

4. $\left(36^{-\frac{1}{4}} \cdot 9^{\frac{3}{4}}\right)^{-2}$

3. _____

4. _____

5. $\left(a^{\frac{5}{4}} \cdot b^{\frac{3}{2}}\right)^{\frac{2}{5}}$

6. $\left(a^{\frac{4}{3}} \cdot b^{-\frac{1}{2}}\right)\left(a^{-2} \cdot b^{\frac{5}{2}}\right)$

5. _____

6. _____

7. $\left(25 p^{\frac{2}{5}} q^{-1}\right)^{\frac{1}{2}}$

8. $\left(\dfrac{64 m^{\frac{1}{2}} n}{m^{-2} n^{\frac{4}{3}}}\right)^{\frac{1}{2}}$

7. _____

8. _____

9. $\left(\dfrac{27 x^{\frac{1}{2}} y^{-1}}{y^{-\frac{2}{3}} x^{-\frac{1}{2}}}\right)^{\frac{1}{3}} - \left(\dfrac{4 x^{\frac{1}{3}} y^{\frac{4}{9}}}{x^{-\frac{1}{3}} y^{\frac{2}{3}}}\right)^{\frac{1}{2}}$

9. _____

In Problems 10 –12, distribute and simplify.

10. $x^{\frac{1}{3}}\left(x^{\frac{5}{3}} + 4\right)$

11. $3a^{-\frac{1}{2}}(2 - a)$

10. _____

11. _____

12. $8 p^{\frac{2}{3}}\left(p^{\frac{4}{3}} - 4 p^{-\frac{2}{3}}\right)$

12. _____

In Problems 13 – 17, use rational exponents to simplify each radical. Assume all variables are positive.

13. $\sqrt[3]{x^6}$

14. $\sqrt[9]{125^6}$

13. _____

14. _____

15. $\sqrt{25x^4y^6}$

16. $\sqrt[4]{p^3} \cdot \sqrt[3]{p}$

15. _____

16. _____

17. $\sqrt{5} \cdot \sqrt[3]{25}$

17. _____

In Problems 18 – 21, simplify by factoring out the given expression.

18. $3(x-5)^{\frac{1}{2}}(3x+1) + 6(x-5)^{\frac{3}{2}}; \ (x-5)^{\frac{1}{2}}$

18. _____

19. $x^{-\frac{2}{3}}(3x+2) + 9x^{\frac{1}{3}}; \ x^{-\frac{2}{3}}$

19. _____

20. $4(x+3)^{\frac{1}{2}} + (x+3)^{-\frac{1}{2}}(2x+1); \ (x+3)^{-\frac{1}{2}}$

20. _____

21. $24x(x^2-1)^{\frac{1}{3}} + 9(x^2-1)^{\frac{4}{3}}; \ (x^2-1)^{\frac{1}{3}}$

21. _____

In Problems 22 – 25, simplify each expression.

22. $\sqrt[6]{27^2}$

23. $25^{\frac{3}{4}} \cdot 25^{\frac{3}{4}}$

22. _____

23. _____

24. $\left(8^4\right)^{\frac{5}{12}}$

25. $\sqrt[9]{a^6} - \dfrac{\sqrt[6]{a^5}}{\sqrt[6]{a}}$

24. _____

25. _____

Sullivan/Struve, *Intermediate Algebra*, 3e
Copyright © 2014 Pearson Education, Inc.

Five-Minute Warm-Up 6.3
Simplifying Radical Expressions Using Properties of Radicals

1. List the perfect squares that are less than 150.   _____

2. List the perfect cubes that are less than 150.   _____

3. List the perfect fourths that are less than 150.   _____

In Problems 4 – 9, simplify each radical.

4. $\sqrt{-4}$ **5.** $\sqrt{10^2}$ 4. _____

                                                                                             5. _____

6. $\sqrt{x^2}$ **7.** $\sqrt{(4x+1)^2}$ 6. _____

                                                                                             7. _____

8. $\sqrt{4p^2 - 4p + 1}$ **9.** $\sqrt{144 + 25}$ 8. _____

                                                                                             9. _____

Guided Practice 6.3
Simplifying Radical Expressions Using Properties of Radicals

Objective 1: Use the Product Property to Multiply Radical Expressions

1. The Product Property of Radicals states that if $\sqrt[n]{a}$ and $\sqrt[n]{b}$ are real numbers and $n \geq 2$ is an integer, then $\sqrt[n]{a} \cdot \sqrt[n]{b} = \sqrt[n]{ab}$. Use this property to multiply each of the following. *(See textbook Example 1)*

(a) $\sqrt{5} \cdot \sqrt{7}$ (b) $\sqrt{x-5} \cdot \sqrt{x+5}$ (c) $\sqrt[3]{7x} \cdot \sqrt[3]{2x}$ 1a. _____

1b. _____

1c. _____

Objective 2: Use the Product Property to Simplify Radical Expressions

2. A radical expression is *simplified* when _____.

3. Simplify: $\sqrt{75}$ *(See textbook Example 2)*

Step 1: Since the index is 2, we write each factor of the radicand as the product of two factors, one of which is a perfect square.   What perfect square is a factor of 75?   (a) _____

Step 2: Write the radical as the product of two radicals, one of which contains the perfect square. $\sqrt{75} =$ (b) _____

Step 3: Take the square root of each perfect power.   (c) _____

4. Simplify: $\dfrac{-6 + \sqrt{48}}{2}$ *(See textbook Example 4)*

$$\dfrac{-6 + \sqrt{48}}{2} =$$

Step 1: Since the index is 2, write each factor of the radicand as the product of the factors, one of which is a perfect square.   (a) _____

Step 2: Use $\sqrt{a \bullet b} = \sqrt{a} \bullet \sqrt{b}$.   (b) _____

Step 3: Take the square root of each perfect power.   (c) _____

Step 4: Factor out the 2 in the numerator.   (d) _____

Step 5: Divide out the common factor.   (e) _____

5. Explain how to simplify an expression with an index that does not divide evenly into the exponent on the variable in the radicand.

Objective 3: Use the Quotient Property to Simplify Radical Expressions

6. Simplify $\sqrt{\dfrac{12x^2}{121}}$. Assume $x \geq 0$. *(See textbook Example 8)*

6. _____

7. Simplify $\dfrac{\sqrt{72a}}{\sqrt{2a^5}}$. Assume $a > 0$. *(See textbook Example 9)*

7. _____

Objective 4: Multiply Radicals with Unlike Indices

8. Multiply: $\sqrt{6} \cdot \sqrt[4]{48}$ *(See textbook Example 10)*

(a) What is the first step to multiply $\sqrt{6} \cdot \sqrt[4]{48}$?

8a. _____

(b) Determine the LCD of 2 and 4.

8b. _____

(c) Use $a^{\frac{n}{m}} = \left(a^{\frac{1}{m}}\right)^n$ to rewrite each factor:

8c. _____

(d) Use $a^r \cdot b^r = (a \cdot b)^r$:

8d. _____

(e) Multiply:

8e. _____

(f) Write the expression using radicals and simplify:

8f. _____

Do the Math Exercises 6.3
Simplifying Radical Expression Using Properties of Radicals

In Problems 1 – 3, use the Product Rule to multiply. Assume that all variables can be any real number.

1. $\sqrt[4]{6a^2} \cdot \sqrt[4]{7b^2}$

2. $\sqrt{p-5} \cdot \sqrt{p+5}$

1. _____

2. _____

3. $\sqrt[3]{\dfrac{-9x^2}{4}} \cdot \sqrt[3]{\dfrac{4}{3x}}$

3. _____

In Problems 4 – 11, simplify each radical using the Product Property. Assume that all variables can be any real number.

4. $\sqrt[4]{162}$

5. $\sqrt{20a^2}$

4. _____

5. _____

6. $\sqrt[3]{-64p^3}$

7. $4\sqrt{27b}$

6. _____

7. _____

8. $\sqrt{s^9}$

9. $\sqrt[5]{x^{12}}$

8. _____

9. _____

10. $\sqrt[3]{-54q^{12}}$

11. $\sqrt{75x^6 y}$

10. _____

11. _____

In Problems 12 and 13, simplify each expression.

12. $\dfrac{5-\sqrt{100}}{5}$

13. $\dfrac{-6+\sqrt{48}}{8}$

12. _____

13. _____

In Problems 14 – 17, multiply and simplify. Assume that all variables are greater than or equal to zero.

14. $\sqrt{3} \cdot \sqrt{12}$

15. $\sqrt{6x} \cdot \sqrt{30x}$

14._____

15._____

16. $\sqrt[3]{9a} \cdot \sqrt[3]{6a^2}$

17. $\sqrt[3]{16m^2n} \cdot \sqrt[3]{27m^2n}$

16._____

17._____

In Problems 18 – 23, simplify. Assume that all variables are greater than zero.

18. $\sqrt{\dfrac{5}{36}}$

19. $\sqrt[4]{\dfrac{5x^4}{16}}$

18._____

19._____

20. $\sqrt[3]{\dfrac{-27x^9}{64y^{12}}}$

21. $\dfrac{\sqrt[4]{64}}{\sqrt[4]{4}}$

20._____

21._____

22. $\dfrac{\sqrt{54y^5}}{\sqrt{3y}}$

23. $\dfrac{\sqrt[3]{-128x^8}}{\sqrt[3]{2x^{-1}}}$

22._____

23._____

In Problems 24 and 25, multiply and simplify.

24. $\sqrt{2} \cdot \sqrt[3]{7}$

25. $\sqrt[4]{3} \cdot \sqrt[8]{5}$

24._____

25._____

Five-Minute Warm-Up 6.4
Adding, Subtracting, and Multiplying Radical Expressions

1. Add: $\left(2z^3 - 7z^2 + 1\right) + \left(z^3 + 2z^2 - 4z - 1\right)$

1. _____

2. Subtract: $\left(-2a^2b^2 + ab - 3b^2\right) - \left(a^2b^2 + a^2 - 5ab - 2b^2\right)$

2. _____

3. Multiply: $-4x^2\left(2x^2 + 3xy - 5y^2\right)$

3. _____

4. Multiply: $\dfrac{3}{2}x^2\left(\dfrac{4}{27}x^3 - \dfrac{8}{21}x^2 + \dfrac{2}{3}\right)$

4. _____

5. Multiply: $\left(9c + 2\right)\left(2c - 3\right)$

5. _____

6. Multiply: $\left(ab - 2\right)\left(ab + 2\right)$

6. _____

7. Multiply: $\left(7n - 3\right)^2$

7. _____

Guided Practice 6.4
Adding, Subtracting, and Multiplying Radical Expressions

⬤ **Objective 1: Add or Subtract Radical Expressions**

1. Describe the characteristics of *like* radicals.

2. Add or subtract, as indicated. Assume all variables are greater than or equal to zero. *(See textbook Example 1)*

(a) $8\sqrt{x} + \sqrt{x}$

(b) $13\sqrt[3]{3p} - 6\sqrt[3]{3p} + \sqrt[3]{3p}$

2a. _____

2b. _____

3. Simplify the radicals and then perform the indicated operations. Assume all variables are greater than or equal to zero. *(See textbook Examples 2 and 3)*

(a) $3\sqrt{8} - 4\sqrt{32}$

(b) $3n^2\sqrt{54n^4} - 2n\sqrt{150n^6} + 2\sqrt{24n^9}$

3a. _____

3b. _____

Objective 2: Multiply Radical Expressions

4. Multiply and simplify: $\left(9 + 2\sqrt{6}\right)\left(2 - 3\sqrt{2}\right)$ *(See textbook Example 4)*

4. _____

5. Multiply and simplify: $\left(7 - \sqrt{3}\right)\left(7 + \sqrt{3}\right)$ *(See textbook Example 5)*

Step 1: What special products formula can be used to multiply $\left(7 - \sqrt{3}\right)\left(7 + \sqrt{3}\right)$?

5a. _____

Step 2: Use the formula from part (a) to multiply and simplify $\left(7 - \sqrt{3}\right)\left(7 + \sqrt{3}\right)$.

5b. _____

6. Use **Heron's Formula** for finding the area of triangle whose sides are known. Heron's Formula states that the area A of a triangle with sides a, b, and c, is

$$A = \sqrt{s(s-a)(s-b)(s-c)}$$

where

$$s = \frac{1}{2}(a+b+c)$$

Find the area of the shaded region by computing the difference in the areas of the triangles. That is, compute "area of larger triangle minus area of smaller triangle." Write your answer as a radical in simplified form.

6. _____

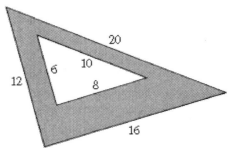

Sullivan/Struve, *Intermediate Algebra*, 3e
Copyright © 2014 Pearson Education, Inc.

Do the Math Exercises 6.4
Adding, Subtracting, and Multiplying Radical Expressions

In Problems 1 – 9, add or subtract as indicated. Assume all variables are positive or zero.

1. $6\sqrt{3} + 8\sqrt{3}$

2. $12\sqrt[4]{z} - 5\sqrt[4]{z}$

1. _____

2. _____

3. $4\sqrt[3]{5} - 3\sqrt{5} + 7\sqrt[3]{5} - 8\sqrt{5}$

4. $6\sqrt{3} + \sqrt{12}$

3. _____

4. _____

5. $7\sqrt[4]{48} - 4\sqrt[4]{243}$

6. $2\sqrt{48z} - \sqrt{75z}$

5. _____

6. _____

7. $3\sqrt{63z^3} + 2z\sqrt{28z}$

8. $\sqrt{48y^2} - 4y\sqrt{12} + \sqrt{108y^2}$

7. _____

8. _____

9. $-2\sqrt[3]{5x^3} + 4x\sqrt[3]{40} - \sqrt[3]{135}$

9. _____

In Problems 10 – 18, multiply and simplify. Assume all variables are positive or zero.

10. $\sqrt{5}\left(5 + 3\sqrt{3}\right)$

11. $\sqrt[3]{6}\left(\sqrt[3]{2} + \sqrt[3]{12}\right)$

10. _____

11. _____

12. $\left(5 + \sqrt{5}\right)\left(3 + \sqrt{6}\right)$

13. $\left(9 + 5\sqrt{10}\right)\left(1 - 3\sqrt{10}\right)$

12. _____

13. _____

14. $\left(\sqrt{6}-2\sqrt{2}\right)\left(2\sqrt{6}+3\sqrt{2}\right)$ **15.** $\left(2-\sqrt{3}\right)^2$ 14. _____

15. _____   ⬤

16. $\left(\sqrt{3}-1\right)\left(\sqrt{3}+1\right)$ **17.** $\left(6+3\sqrt{2}\right)\left(6-3\sqrt{2}\right)$ 16. _____

17. _____

18. $\left(\sqrt[3]{y}-6\right)\left(\sqrt[3]{y}+3\right)$ 18. _____

In Problems 19 – 25, perform the indicated operation and simplify. Assume all variables are positive or zero.

19. $\left(\sqrt{6}-2\sqrt{2}\right)\left(2\sqrt{6}+3\sqrt{2}\right)$ **20.** $\left(\sqrt{7}-\sqrt{3}\right)^2$ 19. _____

20. _____   ⬤

21. $\left(\sqrt{5}-\sqrt{3}\right)^2-\sqrt{60}$ **22.** $\left(4+\sqrt{2x+3}\right)^2$ 21. _____

22. _____

23. $\left(\sqrt{3a}-\sqrt{4b}\right)\left(\sqrt{3a}+\sqrt{4b}\right)+4\sqrt{b^2}$ **24.** $\left(\sqrt{2}-\sqrt{7}\right)^2-\sqrt{56}$ 23. _____

24. _____

25. $\dfrac{4}{5}\cdot\left(-\dfrac{\sqrt{5}}{5}\right)+\left(-\dfrac{3}{5}\right)\cdot\left(-\dfrac{2\sqrt{5}}{5}\right)$ 25. _____

⬤

Sullivan/Struve, *Intermediate Algebra*, 3e
Copyright © 2014 Pearson Education, Inc.

Five-Minute Warm-Up 6.5
Rationalizing Radical Expressions

1. By what would you multiply 75 so that the product is a perfect square? There is more than one right answer so choose the smallest possible factor that yields a perfect square.

1. _____

2. Simplify: $\sqrt{121a^2}$, $a > 0$

2. _____

3. Multiply: $\dfrac{\sqrt{3}}{\sqrt{2}} \cdot \dfrac{\sqrt{6}}{\sqrt{2}}$

3. _____

4. Multiply: $\left(2 + \sqrt{3}\right)\left(2 - \sqrt{3}\right)$

4. _____

5. Multiply: $\left(2\sqrt{5} + \sqrt{3}\right)\left(3\sqrt{5} - 2\sqrt{3}\right)$

5. _____

Guided Practice 6.5
Rationalizing Radical Expressions

Objective 1: Rationalize a Denominator Containing One Term

1. In your own words, what does it mean to rationalize the denominator of a rational expression?

2. To rationalize a denominator containing a single square root, we multiply the numerator and denominator

of the quotient by a square root so that the radicand in the denominator becomes _____.

3. Determine what to multiply each quotient by so that the denominator contains a radicand which is a perfect square. *(See textbook Example 1)*

(a) $\dfrac{3}{\sqrt{3}}$ 　　　　　　**(b)** $\dfrac{4}{\sqrt{20}}$ 　　　　　　**(c)** $\dfrac{1}{\sqrt{8a}}$

3a. _____

3b. _____

3c. _____

4. Determine what to multiply each quotient by so that the denominator contains a radicand which is a perfect power. *(See textbook Example 2)*

(a) $\dfrac{6}{\sqrt[3]{5}}$ 　　　　　　**(b)** $\sqrt[3]{\dfrac{2}{12}}$ 　　　　　　**(c)** $\dfrac{4x}{\sqrt[4]{27x^2y}}$

4a. _____

4b. _____

4c. _____

Objective 2: *Rationalize a Denominator Containing Two Terms*

5. To rationalize a denominator containing two terms, we multiply both numerator and denominator by the

_____ of the denominator.

6. Identify the conjugate of each expression. Then multiply the expression by its conjugate.
(See textbook Example 3)

(a) $3 + \sqrt{5}$ **(b)** $2\sqrt{3} - 5\sqrt{2}$ 6a. _____

                                                                   6b. _____

7. Determine what to multiply the quotient by to rationalize the denominator $\dfrac{\sqrt{8}}{\sqrt{2} - 3}$.   7. _____

(See textbook Example 3)

8. Rationalize the denominator: $\dfrac{12}{\sqrt{2} - \sqrt{5}}$ 8. _____

Do the Math Exercises 6.5
Rationalizing Radical Expressions

In Problems 1 – 14, rationalize each denominator. Assume all variables are positive.

1. $\dfrac{2}{\sqrt{3}}$

2. $-\dfrac{3}{2\sqrt{3}}$

1. _____

2. _____

3. $\dfrac{5}{\sqrt{20}}$

4. $\dfrac{\sqrt{3}}{\sqrt{11}}$

3. _____

4. _____

5. $\sqrt{\dfrac{5}{z}}$

6. $\dfrac{\sqrt{32}}{\sqrt{a^5}}$

5. _____

6. _____

7. $-\sqrt[3]{\dfrac{4}{p}}$

8. $\sqrt[3]{\dfrac{-5}{72}}$

7. _____

8. _____

9. $\dfrac{8}{\sqrt[3]{36z^2}}$

10. $\dfrac{6}{\sqrt[4]{9b^2}}$

9. _____

10. _____

11. $\dfrac{6}{\sqrt{7}-2}$

12. $\dfrac{10}{\sqrt{10}+3}$

11. _____

12. _____

13. $\dfrac{\sqrt{3}}{\sqrt{15}-\sqrt{6}}$

14. $\dfrac{2\sqrt{3}+3}{\sqrt{12}-\sqrt{3}}$

13. _____

14. _____

In Problems 15 – 18, perform the indicated operation and simplify.

15. $\sqrt{5} - \dfrac{1}{\sqrt{5}}$

16. $\dfrac{\sqrt{5}}{2} + \dfrac{3}{\sqrt{5}}$

15. _____

16. _____

17. $\sqrt{\dfrac{2}{5}} + \sqrt{20} - \sqrt{45}$

18. $\sqrt{\dfrac{4}{3}} + \dfrac{4}{\sqrt{48}}$

17. _____

18. _____

In Problems 19 – 23, simplify each expression so that the denominator does not contain a radical.

19. $\dfrac{\sqrt{2}}{\sqrt{18}}$

20. $\sqrt{\dfrac{9}{5}}$

19. _____

20. _____

21. $\dfrac{\sqrt{2} - 5}{\sqrt{2} + 5}$

22. $\dfrac{5}{\sqrt{6} + 4}$

21. _____

22. _____

23. $\dfrac{\sqrt{75}}{\sqrt{3}}$

23. _____

In Problems 24 and 25, rationalize the numerator.

24. $\dfrac{\sqrt{3} + 2}{2}$

25. $\dfrac{\sqrt{a} - \sqrt{b}}{\sqrt{2}}$

24. _____

25. _____

Sullivan/Struve, *Intermediate Algebra*, 3e

Five-Minute Warm-Up 6.6
Functions Involving Radicals

In Problems 1 and 2, simplify each expression.

1. $\sqrt{144}$

2. $\sqrt{n^2}$, $n > 0$

1. _____

2. _____

3. Solve: $-3x - 9 \geq 0$

3. _____

4. Given $f(x) = -2x^2 + 16x$, find $f(-2)$.

4. _____

5. Graph $f(x) = x^2 + 2$ using point plotting.

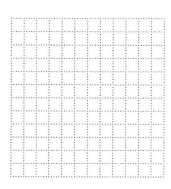

Guided Practice 6.6
Functions Involving Radicals

Objective 1: Evaluate Functions Involving Radicals

1. For the functions $f(x) = 2\sqrt{3x-1}$, $g(x) = \sqrt[3]{\dfrac{3x}{x+4}}$, and $h(x) = \sqrt{\dfrac{x+2}{x-2}}$, find each of the following.

(See textbook Example 1)

(a) $f(7)$ **(b)** $g(-3)$ **(c)** $h(6)$

1a. _____

1b. _____

1c. _____

Objective 2: Find the Domain of a Function Involving a Radical

2. If the index on a radical is even, then the radicand must be _____.

3. If the index on a radical is odd, then the radicand can be _____.

4. Find the domain of each of the following functions. *(See textbook Example 2)*

(a) $f(x) = \sqrt{2x-3}$ **(b)** $g(x) = \sqrt[3]{6x-9}$ **(c)** $h(t) = \sqrt[4]{14-7t}$

4a. _____

4b. _____

4c. _____

Objective 3: Graph Functions Involving Square Roots

5. Given the function $f(x) = \sqrt{x+3}$, *(See textbook Example 3)*

(a) find the domain.

5a. _____

(b) graph the function using point-plotting.

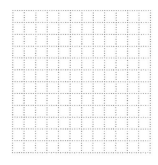

(c) Based on the graph, determine the range.

5c. _____

Objective 4: Graph Functions Involving Cube Roots

6. Given the function $g(x) = \sqrt[3]{x} - 3$, *(See textbook Example 4)*

(a) find the domain.

6a. _____

(b) graph the function using point-plotting.

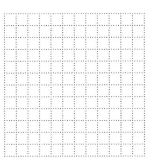

(c) Based on the graph, determine the range.

6c. _____

Sullivan/Struve, *Intermediate Algebra,* 3e
Copyright © 2014 Pearson Education, Inc.

Do the Math Exercises 6.6
Functions Involving Radicals

In Problems 1 – 3, evaluate each radical function at the indicated values.

1. $f(x) = \sqrt{x + 10}$

   (a) $f(6)$

   (b) $f(2)$

   (c) $f(-6)$

2. $G(t) = \sqrt[3]{t - 6}$

   (a) $G(7)$

   (b) $G(-21)$

   (c) $G(22)$

1a. _____

1b. _____

1c. _____

2a. _____

2b. _____

2c. _____

3. $f(x) = \sqrt{\dfrac{x - 4}{x + 4}}$

   (a) $f(5)$

   (b) $f(8)$

   (c) $f(12)$

3a. _____

3b. _____

3c. _____

In Problems 4 – 8, find the domain of the radical function.

4. $f(x) = \sqrt{x + 4}$

5. $G(x) = \sqrt{5 - 2x}$

4. _____

5. _____

6. $G(z) = \sqrt[3]{5z - 3}$

7. $C(y) = \sqrt[4]{3y - 2}$

6. _____

7. _____

8. $f(x) = \sqrt{\dfrac{3}{x - 3}}$

8. _____

In Problems 9 – 12, (a) determine the domain of the function; (b) determine the range of the function; (c) graph the function using point-plotting.

9. $f(x) = \sqrt{x-1}$

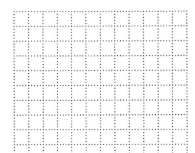

10. $F(x) = \sqrt{4-x}$

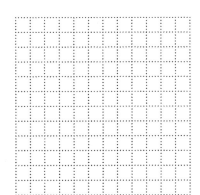

9a. _____

9b. _____

10a. _____

10b. _____

11. $f(x) = \sqrt{x} + 1$

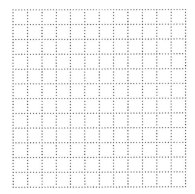

12. $g(x) = \sqrt[3]{x-4}$

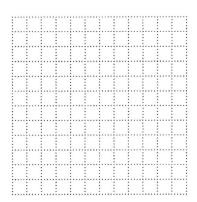

11a. _____

11b. _____

12a. _____

12b. _____

Sullivan/Struve, *Intermediate Algebra*, 3e
Copyright © 2014 Pearson Education, Inc.

Five-Minute Warm-Up 6.7
Radical Equations and Their Applications

1. Solve: $4x + 12 = 0$

1. _____

2. Solve: $-4x^2 + 12x - 8 = 0$

2. _____

3. Simplify: $\left(\sqrt{2x}\right)^2$, $x > 0$

3. _____

4. Multiply: $\left(\sqrt{2x} + 5\right)^2$; $x > 0$

4. _____

5. Simplify: $\left[(3x+8)^{\frac{3}{2}}\right]^{\frac{2}{3}}$; $x > 0$

5. _____

6. Evaluate: $\sqrt{2(-3) + (-3)(-18)}$

6. _____

Guided Practice 6.7
Radical Equations and Their Applications

Objective 1: Solve Radical Equations Containing One Radical

1. Solve: $\sqrt{4x+1} - 2 = 3$ *(See textbook Example 1)*

Step 1: Isolate the radical. $\sqrt{4x+1} - 2 = 3$

                          Add 2 to both sides:

(a) _____

Step 2: Raise both sides to the power of the index.

The index is 2, so we square both sides:

(b) _____

                          Simplify:

(c) _____

Step 3: Solve the equation that results.

(d) _____

Step 4: Check

Substitute the value you found in (d) into the original equation. Does this yield a true statement? **(e)** _____

State the solution set: **(f)** _____

2. What is meant by *extraneous solutions*?

3. When solving radical equations, what is the key first step?

4. Solve: $\sqrt{7x+2} + 5 = 2$ *(See textbook Example 2)*

(a) What value did you find for *x*?   4a. _____

(b) Does this satisfy the original equation?   4b. _____

(c) State the solution set.   4c. _____

(d) Without solving, how can you tell that this equation has no real solution?   4d. _____

5. Solve: $\sqrt{x+5} + 1 = x$ *(See textbook Example 3)*

(a) What value(s) did you find for x?   5a. _____

(b) Does this value (or values) satisfy the original equation?   5b. _____

(c) State the solution set.   5c. _____

6. Solve: $(3x+1)^{\frac{3}{2}} - 2 = 6$ *(See textbook Example 5)*

(a) Isolate the expression containing the rational exponent.   6a. _____

(b) To what power do we need to raise both sides in order to eliminate the exponent on the variable expression?   6b. _____

(c) Solve the resulting equation. What value(s) did you find for x?   6c. _____

(d) Check and then state the solution set.   6d. _____

Objective 2: Solve Radical Equations Containing Two Radicals

7. Solve: $\sqrt{2x^2 - 5x - 20} = \sqrt{x^2 - 3x + 15}$ *(See textbook Example 6)*

		$\sqrt{2x^2 - 5x - 20} = \sqrt{x^2 - 3x + 15}$
Step 1: Isolate one of the radicals.	The radical on the left side of the equation is isolated:	**(a)** _____
Step 2: Raise both sides to the power of the index.	The index is 2, so we square both sides:	**(b)** _____
Step 3: Because there is no radical, we solve the equation that results.	Solve:	**(c)** _____
Step 4: Check.	Substitute the value(s) you found in (b) into the original equation. Does this yield a true statement? State the solution set:	**(d)** _____ **(e)** _____

Do the Math Exercises 6.7
Radical Equations and Their Applications

In Problems 1 – 18, solve each equation.

1. $\sqrt{y-5} = 3$

2. $\sqrt{6p-5} = -5$

1. _____

2. _____

3. $\sqrt[3]{9w} = 3$

4. $\sqrt{q} - 5 = 2$

3. _____

4. _____

5. $\sqrt{x-4} + 4 = 7$

6. $4\sqrt{t} - 2 = 10$

5. _____

6. _____

7. $\sqrt{w} = 6 - w$

8. $\sqrt{1-4x} - 5 = x$

7. _____

8. _____

9. $\sqrt{3x+1} = \sqrt{2x+7}$

10. $\sqrt{2x^2 + 7x - 10} = \sqrt{x^2 + 4x + 8}$

9. _____

10. _____

11. $\sqrt{2x-1} - \sqrt{x-1} = 1$

12. $\sqrt{4x+1} - \sqrt{2x+1} = 2$

11. _____

12. _____

13. $(6p+3)^{1/5} = (4p-9)^{1/5}$

14. $(2x+3)^{1/2} = 3$

13. _____

14. _____

15. $\sqrt{x+20} = x$

16. $\sqrt[5]{x+23} = 2$

15. _____

16. _____

17. $\sqrt{x-3}+\sqrt{x+4}=7$

18. $\sqrt{a}-5=-2$

17. _____

18. _____

In Problems 19 – 21, solve for the indicated variable.

19. Solve $v=\sqrt{ar}$ for a.

20. Solve $r=\sqrt{\dfrac{S}{4\pi}}$ for S.

19. _____

20. _____

21. Solve $V=\sqrt{\dfrac{2U}{C}}$ for U.

21. _____

22. Money The annual rate of interest r (expressed as a decimal) required to have A dollars after t years from an initial deposit of P dollars can be calculated with the following formula:

$$r=\sqrt[t]{\frac{A}{P}}-1$$

Suppose that you deposit \$1,000 in an account that pays 5% annual interest so that $r = 0.05$. How much will you have after $t = 2$ years?

22. _____

Five-Minute Warm-Up 6.8
The Complex Number System

1. List the numbers in the set $\left\{ 8, \dfrac{-6}{3}, \left| -5 \right|, 0.\overline{3}, \dfrac{15}{0}, -\dfrac{2}{5}, \pi, \dfrac{0}{-12} \right\}$ which are:

(a) Natural numbers

1a. _____

(b) Whole numbers

1b. _____

(c) Integers

1c. _____

(d) Rational numbers

1d. _____

(e) Irrational numbers

1e. _____

(f) Real numbers

1f. _____

In Problems 2 – 5, perform the indicated operations and simplify.

2. $\left(x^2 - 1 \right) + \left(2x^2 - 1 \right)$

2. _____

3. $4p^5 \left(-3p + 9 \right)$

3. _____

4. $\left(2x - 7 \right)\left(-3x - 4 \right)$

4. _____

5. $\left(x^2 + 4 \right)\left(x^2 - 4 \right)$

5. _____

Guided Practice 6.8
The Complex Number System

1. The *imaginary unit*, denoted by *i,* is the number whose square is -1. $i^2 =$ _____ or $i =$ _____.

2. *Complex numbers* are numbers of the form _____ where *a* and *b* are _____. When a number is in a form such as $6 - 2i$, we say that the number is in _____ form. The *real part* of the complex number is _____ and the *imaginary part* is _____.

Objective 1: Evaluate the Square Root of Negative Real Numbers

3. Write $\sqrt{-81}$ as a pure imaginary number. *(See textbook Example 1)*

3. _____

4. Write $6 + \sqrt{-4}$ in standard form. *(See textbook Example 2)*

4. _____

Objective 2: Add or Subtract Complex Numbers

5. Before beginning any operations with complex numbers, you must write the number in _____ form.

6. In your own words, explain how to add complex numbers _____

_____

Objective 3: Multiply Complex Numbers

7. Use the Distributive Property to multiply $\frac{3}{2}i(6 - 8i)$. *(See textbook Example 5)*

7. _____

8. Multiply $\sqrt{-4} \cdot \sqrt{-9}$. *(See textbook Example 6)*
(a) Explain why the Product Property of Radicals cannot be used to multiply these radicals.

(a)_____

(b) Express the radicals as pure imaginary numbers.

8b._____

(c) Multiply.

8c._____

Objective 4: Divide Complex Numbers

9. Divide: $\dfrac{3 + 6i}{4 + 4i}$. *(See textbook Example 8)*

Step 1: Write the numerator and denominator in standard form, *a + bi*.

The numerator and denominator are already in standard form.

Step 2: Multiply the numerator and denominator by the complex conjugate of the denominator.

Identify the conjugate of the denominator:

(a) _____

Multiply the quotient by 1, written with the conjugate:

(b) _____

Step 3: Simplify by writing the quotient in standard form, *a + bi*.

Multiply the numerator; the denominator is of the form $(a + bi)(a - bi) = a^2 + b^2$:

(c) _____

Combine like terms; $i^2 = -1$:

(d) _____

Divide the denominator into each term of the numerator to write in standard form:

(e) _____

Write each fraction in lowest terms:

(f) _____

Objective 5: Evaluate the Powers of *i*

10. The powers of *i* are a cyclic function, meaning that the values cycle through a set list. Complete the table to see the only values for the powers of *i*.

$i^1 = i$

(a) $i^2 =$ _____

(b) $i^3 =$ _____

(c) $i^4 =$ _____

(d) $i^5 =$ _____

(e) $i^6 =$ _____

(f) $i^7 =$ _____

(g) $i^8 =$ _____

Do the Math Exercises 6.8
The Complex Number System

In Problems 1 and 2, write each expression as a pure imaginary number.

1. $-\sqrt{-100}$ **2.** $\sqrt{-162}$

1. _____

2. _____

In Problems 3 – 5, write each expression as a complex number in standard form.

3. $10 + \sqrt{-32}$ **4.** $\dfrac{10 - \sqrt{-25}}{5}$

3. _____

4. _____

5. $\dfrac{15 - \sqrt{-50}}{5}$

5. _____

In Problems 6 – 9, add or subtract as indicated.

6. $(-6 + 2i) + (3 + 12i)$ **7.** $(-7 + 3i) - (-3 + 2i)$

6. _____

7. _____

8. $\left(-4 + \sqrt{-25}\right) + \left(1 - \sqrt{-16}\right)$ **9.** $\left(-10 + \sqrt{-20}\right) - \left(-6 + \sqrt{-45}\right)$

8. _____

9. _____

In Problems 10 – 17, multiply.

10. $3i(-2 - 6i)$ **11.** $(3 - i)(1 + 2i)$

10. _____

11. _____

12. $(2 + 8i)(-3 - i)$ **13.** $\left(-\dfrac{2}{3} + \dfrac{4}{3}i\right)\left(\dfrac{1}{2} - \dfrac{3}{2}i\right)$

12. _____

13. _____

14. $(2 + 5i)^2$

15. $(2 - 7i)^2$

14. _____

15. _____

16. $\sqrt{-36} \cdot \sqrt{-4}$

17. $\left(1 - \sqrt{-64}\right)\left(-2 + \sqrt{-49}\right)$

16. _____

17. _____

In Problems 18 – 21, divide.

18. $-\dfrac{2 - i}{2i}$

19. $\dfrac{2}{4 + i}$

18. _____

19. _____

20. $\dfrac{-4}{-5 - 3i}$

21. $\dfrac{2 + 5i}{5 - 2i}$

20. _____

21. _____

In Problems 22 – 24, simplify.

22. i^{72}

23. i^{110}

22. _____

23. _____

24. i^{131}

24. _____

Sullivan/Struve, *Intermediate Algebra*, 3e
Copyright © 2014 Pearson Education, Inc.

Five-Minute Warm-Up 7.1
Solving Quadratic Equations by Completing the Square

1. Multiply: $(3x - 1)^2$

1. _____

2. Factor: $x^2 - 4x + 4$

2. _____

In Problems 3 and 4, solve each polynomial equation.

3. $x^2 - \dfrac{2}{3}x + \dfrac{1}{9} = 0$

4. $25n^2 - 49 = 0$

3. _____

4. _____

In Problems 5 and 6, simplify each expression.

5. $\sqrt{\dfrac{81}{25}}$

6. $\sqrt{(8x - 3)^2}$

5. _____

6. _____

In Problems 7 and 8, simplify each expression using complex numbers.

7. $\sqrt{-12}$

8. $\dfrac{8 + \sqrt{-4}}{4}$

7. _____

8. _____

9. Find the complex conjugate of $-15 + 7i$.

9. _____

Name: Date:

Instructor: Section:

Guided Practice 7.1
Solving Quadratic Equations by Completing the Square

Objective 1: Solve Quadratic Equations Using the Square Root Property

1. State the Square Root Property: If $x^2 = p$, where x is any variable expression and p is a real number,

then _____.

2. If the solution to a quadratic equation is $n = -1 \pm 2\sqrt{3}$, write the solution set. _____

3. *True or False* $\sqrt{x^2 - 16} = \sqrt{81}$ simplifies to $x - 4 = \pm 9$. _____

4. *True or False* $y^2 = -9$ has no solution. _____

5. Solve: $n^2 - 144 = 0$ *(See textbook Example 1)*

Step 1: Isolate the expression containing the square term.

$$n^2 - 144 = 0$$

Add 144 to both sides: **(a)** _____

Step 2: Use the Square Root Property. Don't forget the $\pm$ symbol.

Take the square root of both sides of the equation: **(b)** _____

Simplify the radical: **(c)** _____

Step 3: Isolate the variable, if necessary.

The variable is already isolated.

Step 4: Verify your solution(s).

State the solution set: **(d)** _____

6. There is no reason that the solution to a quadratic equation must be real. List the possible nature of the solution(s) of a quadratic equation.

(a) Real, which includes: _____, _____, _____ and

(b) _____ in form _____

Objective 2: Complete the Square in One Variable

7. If a polynomial is of the form $x^2 + bx + c$, c must be equal to _____ in order to be a perfect square trinomial.

8. Determine the number that must be added to the expression to make it a perfect square trinomial. Then factor the expression. *(See textbook Example 5)*

(a) $p^2 - 14p$ **(b)** $n^2 + 9n$ **(c)** $z^2 + \dfrac{4}{3}z$

8a. _____

8b. _____

8c. _____

Objective 3: Solve Quadratic Equations by Completing the Square

9. Solve: $p^2 - 6p - 18 = 0$ *(See textbook Example 6)*

Step 1: Rewrite $x^2 + bx + c = 0$ as $x^2 + bx = -c$ by adding or subtracting the constant from both sides of the equation.

Add 18 to both sides:

$p^2 - 6p - 18 = 0$

(a) _____

Step 2: Complete the square in the expression $x^2 + bx$ by making it a perfect square trinomial.

What value must be added to both sides to make the expression on the left a perfect square trinomial?

(b) _____

Add this number to both sides of the equation and simplify:

(c) _____

Step 3: Factor the perfect square trinomial on the left side of the equation.

Use: $A^2 - 2AB + B^2 = (A - B)^2$

(d) _____

Step 4: Solve the equation using the Square Root Property.

Take the square root of both sides of the equation:

(e) _____

Simplify the square root:

(f) _____

Add 3 to both sides:

(g) _____

Step 5: Verify your solutions(s).

State the solution set:

(h) _____

10. To solve the equation $3x^2 - 9x + 12 = 0$ by the completing the square, the first step is_____.

Objective 4: Solve Problems Using the Pythagorean Theorem

11. State the Pythagorean Theorem in words. If x and y are the lengths of the legs and z is the length of the hypotenuse, write an equation which uses these variables to state the Pythagorean Theorem.

Do the Math Exercises 7.1
Solving Quadratic Equations by Completing the Square

In Problems 1 – 6, solve each equation using the Square Root Property.

1. $z^2 = 48$ **2.** $w^2 - 6 = 14$

1. _____

2. _____

3. $(y - 2)^2 = 9$ **4.** $(2p + 3)^2 = 16$

3. _____

4. _____

5. $\left(y + \dfrac{3}{2}\right)^2 = \dfrac{3}{4}$ **6.** $q^2 - 6q + 9 = 16$

5. _____

6. _____

In Problems 7 – 9, find the value to complete the square in each expression. Then factor the perfect square trinomial.

7. $p^2 - 4p$ **8.** $z^2 - \dfrac{1}{3}z$

7. _____
_____

8. _____
_____

9. $m^2 + \dfrac{5}{2}m$

9. _____
_____

In Problems 10 – 15, solve each quadratic equation by completing the square.

10. $y^2 + 3y - 18 = 0$ **11.** $q^2 + 7q + 7 = 0$

10. _____

11. _____

12. $x^2 - 5x - 3 = 0$ **13.** $n^2 = 10n + 5$

12. _____

13. _____

14. $3a^2 - 4a - 4 = 0$ **15.** $2z^2 + 6z + 5 = 0$

14. _____

15. _____

In Problems 16 and 17, the lengths of the legs of a right triangle are given. Find the length of the hypotenuse. Give the exact answers and decimal approximations rounded to two decimal places.

16. $a = 7, b = 24$ **17.** $a = 2, b = \sqrt{5}$

16. _____

17. _____

18. Right Triangle A right triangle has a leg of length 2 and hypotenuse of length 10. Find the length of the missing leg. Give the exact answer and a decimal approximation, rounded to 2 decimal places.

18. _____

19. Given that $h(x) = (x + 1)^2$, find all values of x such that $h(x) = 32$.

19. _____

20. Fire Truck Ladder A fire truck has a 75-foot ladder. If the truck can safely park 20 feet from a building, how far up the building can the ladder reach assuming that the top of the base of the ladder is resting on top of the truck and the truck is 10 feet tall? Give the decimal approximation, rounded to 3 decimal places.

20. _____

21. The converse of the Pythagorean Theorem is also true. That is, in a triangle, if the square of the length of one side equals the sum of the squares of the lengths of the other two sides, then the triangle is a right triangle. The 90° angle is opposite the longest side.

The lengths of the sides of a triangle are: 20, 48, and 52. Determine whether the triangle is a right triangle. If it is, identify the hypotenuse.

21. _____

Sullivan/Struve, *Intermediate Algebra*, 3e
Copyright © 2014 Pearson Education, Inc.

Five-Minute Warm-Up 7.2
Solving Quadratic Equations by the Quadratic Formula

In Problems 1 and 2, simplify each expression.

1. $\sqrt{125}$

2. $\dfrac{15 + \sqrt{72}}{6}$

1. _____

2. _____

In Problems 3 and 4, simplify each expression using complex numbers.

3. $\sqrt{-28}$

4. $\dfrac{6 - 2\sqrt{-4}}{6}$

3. _____

4. _____

5. Divide: $\dfrac{9x^2 - 27x + 3}{3}$

5. _____

6. Evaluate the expression $\sqrt{b^2 - 4ac}$ if $a = 2,\, b = -4,\, c = -3$.

6. _____

Guided Practice 7.2
Solving Quadratic Equations by the Quadratic Formula

● **Objective 1: Solve Quadratic Equations Using the Quadratic Formula**

1. If $ax^2 + bx + c = 0$, then $x =$ _____.

2. When using the quadratic formula, the first step is to write the quadratic equation in _____.

3. Write the quadratic equation in standard form and then identify the values assigned to a, b, and c.
   Do not solve the equation.
   (a) $x - 2x^2 = -4$ **(b)** $2 - 3x^2 = 8$ **(c)** $3x^2 = 6x$ 3a. _____

   3b. _____

   3c. _____

4. Solve: $8n^2 - 2n - 3 = 0$ *(See textbook Example 1)*

   Step 1: Write the equation in standard form $ax^2 + bx + c = 0$ and identify the values of a, b, and c.

   Since the equation is already in standard form, identify the values for a, b, and c.

   $8n^2 - 2n - 3 = 0$

   (a) $a =$ _____; $b =$ _____; $c =$ _____

   Step 2: Substitute the values of a, b, and c into the quadratic formula.

   Write the quadratic formula: **(b)** _____

   Substitute the values for a, b, and c.   **(c)** _____

   Step 3: Simplify the expression found in Step 2.

   What is the value of the radicand?   **(d)** _____

   Simplify the expression in (c):   **(e)** _____

   Write the two expressions using $a \pm b$ means $a - b$ or $a + b$:   **(f)** _____

   Step 4: Check

   State the solution set: **(g)** _____

Objective 2: Use the Discriminant to Determine the Nature of Solutions of a Quadratic Equation

5. In the quadratic equation $ax^2 + bx + c = 0$, the *discriminant* is used to determine the nature and number of solutions. To find the discriminant, substitute the identified values for a, b, and c into part of the quadratic

formula, _____.

6. Determine the discriminant of each quadratic equation. Use the value of the discriminant to determine whether the quadratic equation has two rational solutions, two irrational solutions, one repeated real solution, or two complex solutions that are not real. *(See textbook Example 5)*

6a. _____

(a) $4x^2 + 5x - 9 = 0$ **(b)** $x^2 + 4x + 9 = 0$ **(c)** $4x^2 = 4x - 1$

_____

6b. _____

_____

6c. _____

_____

Objective 3: Model and Solve Problems Involving Quadratic Equations

7. Projectile Motion The height s of a toy rocket after t seconds, when fired straight up with an initial speed of 150 feet per second from an initial height of 2 feet, can be modeled by the function

$$s(t) = -16t^2 + 150t + 2$$

When will the height of the rocket be 200 feet? Round your answer to the nearest tenth of a second. *(See textbook Examples 6 and 7)*

(a) Step 1: Identify Here we want to know the value t when $s = ?$ _____

    Step 2: Name The variables are named in the problem. t is the time the rocket has traveled and s is height of the rocket.

(b) Step 3: Translate Use the information from (a) and the given formula to write a model for this problem.

_____

(c) Step 4: Solve Solve the equation in (b). What values did you find for t? _____

    Step 5: Check

(d) Step 6: Answer the Question _____

(e) Will the rocket ever reach a height of 500 feet? _____

(f) When will the rocket hit the ground? _____

Sullivan/Struve, *Intermediate Algebra*, 3e
Copyright © 2014 Pearson Education, Inc.

Do the Math Exercises 7.2
Solving Quadratic Equations by the Quadratic Formula

In Problems 1 – 6, solve each equation using the quadratic formula.

1. $p^2 - 4p - 32 = 0$

2. $10x^2 + x - 2 = 0$

1. _____

2. _____

3. $2q^2 - 4q + 1 = 0$

4. $x + \dfrac{1}{x} = 3$

3. _____

4. _____

5. $2z^2 + 7 = 4z$

6. $1 = 5w^2 + 6w$

5. _____

6. _____

In Problems 7 – 9, determine the discriminant of each quadratic equation. Use the value of the discriminant to determine whether the quadratic equation has two rational solutions, two irrational solutions, one repeated real solution, or two complex solutions that are not real.

7. $p^2 + 4p - 2 = 0$

8. $16x^2 + 24x + 9 = 0$

7. _____

8. _____

9. $6x^2 - x = -4$

9. _____

In Problems 10 – 15, solve each equation using any method you wish.

10. $q^2 - 7q + 7 = 0$

11. $3x^2 + 5x = 2$

10. _____

11. _____

12. $5m - 4 = \dfrac{5}{m}$

13. $8p^2 - 40p + 50 = 0$

12. _____

13. _____

14. $(a - 3)(a + 1) = 2$

15. $\dfrac{x - 1}{x^2 + 4} = 1$

14. _____

15. _____

In Problem 16, suppose that $f(x) = x^2 + 2x - 8$.

16a. Solve for x, if $f(x) = 0$

16b. Solve for x, if $f(x) = -8$

16a. _____

16b. _____

17. Area The area of a rectangle is 60 square inches. The length of the rectangle is 6 inches more than the width. What are the dimensions of the rectangle?

17. _____

18. Area The area of a triangle is 35 square inches. The height of the rectangle is 2 inches less than the base. What are the base and height of the triangle?

18. _____

19. Roundtrip A Cessna aircraft flies 200 miles due west into the jet stream and flies back home on the same route. The total time of the trip (excluding the time on the ground) takes 4 hours. The Cessna aircraft can fly 120 miles per hour in still air. What is the net effect of the jet stream on the aircraft?

19. _____

20. Given the quadratic equation $ax^2 + bx + c = 0$, show that the sum of the solutions to any quadratic equation in this form is $-\dfrac{b}{a}$. Show that the product of the solutions to this quadratic equation is $\dfrac{c}{a}$.

Five-Minute Warm-Up 7.3
Solving Equations Quadratic in Form

In Problems 1 and 2, factor completely.

1. $a^4 - 7a^2 - 18$

2. $6(x-1)^2 - 13(x-1) + 6$

1. _____

2. _____

In Problems 3 and 4, simplify each expression.

3. $\left(5x^{-1}\right)^2$

4. $\left(-\dfrac{2}{3}x^3\right)^2$

3. _____

4. _____

5. Solve: $2x^2 + x - 6 = 0$

5. _____

Guided Practice 7.3
Solving Equations Quadratic in Form

Objective 1: Solve Equations That Are Quadratic in Form

1. In general, if a substitution u transforms an equation into one of the form $au^2 + bu + c = 0$, then the original equation is called an **equation quadratic in form.** Here u represents any variable expression and we solve the equation by substituting u for the variable expression written with the coefficient b. For these equations quadratic in form, identify the substitution for u and then write a new equation using your substitution.

(a) $2y - 11\sqrt{y} + 15 = 0$ $u = $ _____   _____

(b) $v^4 + 10v^2 + 1 = 0$ $u = $ _____   _____

(c) $(x-1)^2 - 5(x-1) + 6 = 0$ $u = $ _____   _____

(d) $x^{\frac{2}{3}} - x^{\frac{1}{3}} - 3 = 0$ $u = $ _____   _____

(e) $4x^{-2} + x^{-1} - 3 = 0$ $u = $ _____   _____

2. Solve: $x^4 - 6x^2 - 16 = 0$ *(See textbook Example 1)*

Step 1: Determine the appropriate substitution and write the equation in the form $au^2 + bu + c = 0$.

Here we let $u = ?$ **(a)** _____

Rewrite the equation using your substitution from (a):

$$x^4 - 6x^2 - 16 = 0$$

    (b) _____

Step 2: Solve the equation $au^2 + bu + c = 0$.

Factor: **(c)** _____

Set each factor to 0 and solve:

    (d) _____

Step 3: Solve for the variable in the original equation using the value of u from (a).

Substitute from (a) into your solution from (d):   **(e)** _____

Solve the equations. In this case we use the Square Root Property: **(f)** _____

Simplify the radicals: **(g)** _____

Step 4: Verify your solution(s).

State the solution set: **(h)** _____

3. Solve: $\left(x^2 + 3\right)^2 - 6\left(x^2 + 3\right) + 8 = 0$ *(See textbook Example 2)*

(a) Here we let $u = $ _____.

(b) In this problem, $u = 4$ or $u = 2$. Replace u and solve for x. State the solution set. _____

4. Solve: $2x + \sqrt{x} - 10 = 0$ *(See textbook Example 3)*

(a) Here we let $u = $ _____.

(b) In this problem, $u = -\dfrac{5}{2}$ or $u = 2$. Replace u and solve for x. State the solution set. _____

5. Solve: $n^{\frac{2}{3}} - 2n^{\frac{1}{3}} - 3 = 0$ *(See textbook Example 5)*

(a) Here we let $u = $ _____.

(b) In this problem, $u = 3$ or $u = -1$. Replace u and solve for x. State the solution set. _____

6. As always, it is important to check for extraneous solutions. List two cases when extraneous solutions might appear.

(a) _____   **(b)** _____

Sullivan/Struve, *Intermediate Algebra*, 3e
Copyright © 2014 Pearson Education, Inc.

Do the Math Exercises 7.3
Solving Equations Quadratic in Form

In Problems 1 – 14, solve each equation.

1. $x^4 - 10x^2 + 9 = 0$

2. $4b^4 - 5b^2 + 1 = 0$

1. _____

2. _____

3. $(x+2)^2 - 3(x+2) - 10 = 0$

4. $x - 5\sqrt{x} - 6 = 0$

3. _____

4. _____

5. $z + 7\sqrt{z} + 6 = 0$

6. $q^{-2} + 2q^{-1} = 15$

5. _____

6. _____

7. $10a^{-2} + 23a^{-1} = 5$

8. $y^{\frac{2}{3}} - 2y^{\frac{1}{3}} - 3 = 0$

7. _____

8. _____

9. $\dfrac{1}{x^2} - \dfrac{7}{x} + 12 = 0$

10. $y^6 - 7y^3 - 8 = 0$

9. _____

10. _____

11. $6b^{-2} - b^{-1} = 1$

12. $x^4 + 3x^2 = 4$

11. _____

12. _____

13. $c^{\frac{1}{2}} + c^{\frac{1}{4}} - 12 = 0$

14. $\left(\dfrac{1}{x-1}\right)^2 + \dfrac{7}{x-1} = 8$

13. _____

14. _____

In Problems 15 and 16, suppose that $f(x) = x^4 + 5x^2 + 3$. Find the values of x such that

15. $f(x) = 3$

16. $f(x) = 17$

15. _____

16. _____

In Problems 17 and 18, suppose that $h(x) = 3x^4 - 9x^2 - 8$. Find the values of x such that

17. $h(x) = -8$

18. $h(x) = 22$

17. _____

18. _____

In Problems 19 and 20, find the zeros of the function.

19. $f(x) = x^4 - 13x^2 + 42$

20. $h(p) = 8p - 18\sqrt{p} - 35$

19. _____

20. _____

Sullivan/Struve, *Intermediate Algebra*, 3e
Copyright © 2014 Pearson Education, Inc.

Five-Minute Warm-Up 7.4
Graphing Quadratic Functions Using Transformations

In Problems 1 and 2, graph using the point-plotting method.

1. $y = -x^2$

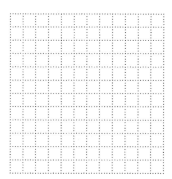

2. $y = x^2 + 2$

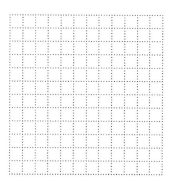

In Problems 3 and 4, find the function value.

3. $f(x) = -x^2 + 6x - 2$; $f(-3)$

4. $f(x) = 2x^2 + x + 5$; $f(-4)$

3._____

4._____

5. What is the domain of $f(x) = \frac{1}{2}x^2 + \frac{4}{3}x - 6$?

5._____

Guided Practice 7.4
Graphing Quadratic Functions Using Transformations

● **Objective 1: Graph Quadratic Functions of the Form** $f(x) = x^2 + k$

In Problems 1 and 2, match the function to either graph (a) *or graph* (b).

1. $f(x) = x^2$ **(a)**   **(b)**   1. _____

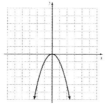 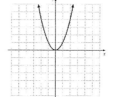

2. $f(x) = -x^2$ 2. _____

3. To obtain the graph of $f(x) = x^2 - 2$, shift the graph of $f(x) = x^2$

(a) horizontally or vertically? *(See textbook Examples 1 and 2)*   3a. _____

(b) How many units will the graph shift?   3b. _____

(c) Does the graph shift right, left, up, or down?   3c. _____

● **Objective 2: Graph Quadratic Functions of the Form** $f(x) = (x - h)^2$

4. To obtain the graph of $f(x) = -(x - 1)^2$, shift the graph of $f(x) = -x^2$

(a) horizontally or vertically? *(See textbook Examples 3 and 4)*   4a. _____

(b) How many units will the graph shift?   4b. _____

(c) Does the graph shift right, left, up, or down?   4c. _____

Objective 3: Graph Quadratic Functions of the Form $f(x) = ax^2$

5. The graph of $f(x) = ax^2 + bx + c$ is a parabola that opens either up or down. This is determined by the coefficient a. *(See textbook Example 6)*
(a) If $a > 0$, the parabola opens   5a. _____

(b) If $a < 0$, the parabola opens   5b. _____

6. The value of a will also determine the breadth of the parabola.

● **(a)** If $|a|$ _____ we say the graph is vertically stretched (taller, thinner, steeper).

(b) If _____ $|a|$ _____ we say the graph is vertically compressed (shorter, fatter, flatter).

Objective 4: Graph Quadratic Functions of the Form $f(x) = ax^2 + bx + c$

7. Graph $f(x) = 3x^2 - 12x + 7$. *(See textbook Examples 7 and 8)*

Step 1: Write the function
$f(x) = ax^2 + bx + c$ as
$f(x) = a(x - h)^2 + k$ by
completing the square.

$f(x) = 3x^2 - 12x + 7$

Group the terms involving x: **(a)** _____

Factor out the coefficient of the **(b)** _____
square term, 3, from the
parentheses:

Identify the number required to **(c)** _____
complete the square:

When this added to the right side of
the equation, what must be done to **(d)** _____
maintain the equality?

Write the amended equation: **(e)** _____

Factor the perfect square trinomial: **(f)** _____

Step 2: Graph the function   Identify the vertex: **(g)** _____
$f(x) = a(x - h)^2 + k$ using
transformations.   Opening direction: **(h)** _____

                                 Axis of symmetry: **(i)** _____

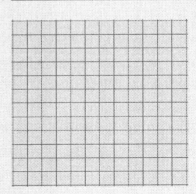

Sullivan/Struve, *Intermediate Algebra*, 3e
Copyright © 2014 Pearson Education, Inc.

Do the Math Exercises 7.4
Graphing Quadratic Functions Using Transformations

In Problems 1 – 6, use the graph of $y = x^2$ to graph the quadratic function.

1. $f(x) = x^2 - 1$

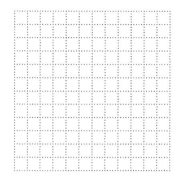

2. $f(x) = (x + 4)^2$

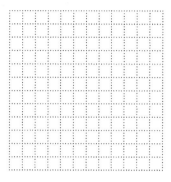

3. $G(x) = 5x^2$

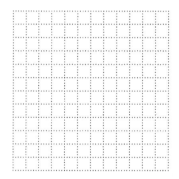

4. $P(x) = -3x^2$

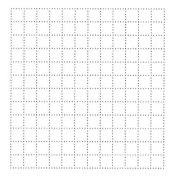

5. $g(x) = (x + 2)^2 - 1$

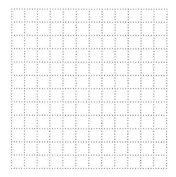

6. $G(x) = (x - 4)^2 + 2$

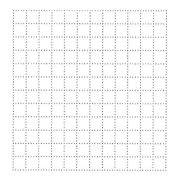

In Problems 7 – 10, write each function in the form $f(x) = a(x - h)^2 + k$. *Then determine the vertex and the axis of symmetry.*

7. $h(x) = x^2 - 7x + 10$

8. $f(x) = 3x^2 + 18x + 25$

7. _____

8. _____

9. $g(x) = -x^2 - 8x - 14$

10. $h(x) = -4x^2 + 4x$

9. _____

10. _____

11. Write a quadratic function in the form $f(x) = a(x - h)^2 + k$ with the properties: opens up; vertically compressed by a factor of $\frac{1}{2}$; vertex at $(-5, 0)$.

11. _____

12. Determine the quadratic function whose graph is shown below. Each tick mark represents one unit of length.

12. _____

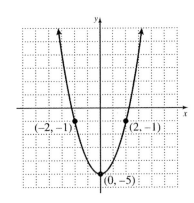

Sullivan/Struve, *Intermediate Algebra*, 3e
Copyright © 2014 Pearson Education, Inc.

Five-Minute Warm-Up 7.5
Graphing Quadratic Functions Using Properties

1. Find the intercepts of the graph of $3x - 4y = -12$.

1. _____

                 _____

2. Solve: $2x^2 - 13x - 24 = 0$

2. _____

3. Find the zeros of $f(x) = -2x^2 + 10x + 12$

3. _____

4. If $f(x) = -x^2 - 3x - 2$, find $f(-5)$.

4. _____

Guided Practice 7.5
Graphing Quadratic Functions Using Properties

Objective 1: Graph Quadratic Functions of the Form $f(x) = ax^2 + bx + c$

1. The vertex is the turning point of the parabola. If $f(x) = ax^2 + bx + c$, the x-coordinate of the vertex is at

$x = $ _____.

2. The x-intercepts of the parabola can be found by _____.

3. Graph $f(x) = x^2 + 2x - 8$ using its properties. *(See textbook Example 1)*

Step 1: Determine whether the parabola opens up or down.

Determine the values using $f(x) = ax^2 + bx + c$:

(a) $a = $ _____; $b = $ _____; $c = $ _____

Does the parabola open up or down?

(b) _____

Step 2: Determine the vertex and axis of symmetry.

Calculate the x-coordinate of the vertex $\left(x = -\dfrac{b}{2a} \right)$:

(c) _____

Calculate the y-coordinate of the vertex:

(d) _____

Identify the vertex:

(e) _____

Identify the axis of symmetry:

(f) _____

Step 3: Determine the y-intercept.

Evaluate $f(0)$:

(g) _____

Step 4: Find the discriminant, $b^2 - 4ac$, to determine the number of the x-intercepts. Then determine the x-intercepts, if any.

Evaluate the discriminant:

(h) _____

Number of x-intercepts:

(i) _____

Factor $0 = x^2 + 2x - 8$ and use the Zero-Product Property to identify the x-intercepts:

(j) _____

8. $g(x) = (x + 2)^2 - 1$

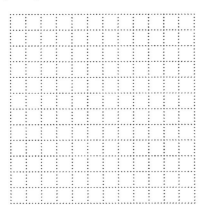

9. $G(x) = (x - 4)^2 + 2$

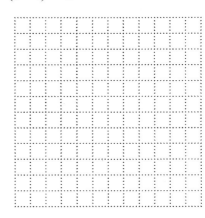

In Problems 10 and 11, determine whether the quadratic function has a maximum or minimum. Then find the maximum or minimum value.

10. $H(x) = -3x^2 + 12x - 1$

11. $G(x) = 5x^2 + 10x - 1$

10. _____

11. _____

12. Fun with Numbers The sum of two numbers is 50. Find the numbers such that their product is a maximum.

12. _____

13. Fun with Numbers The difference of two numbers is 10. Find the numbers such that their product is a minimum.

13. _____

14. Punkin Chunkin Suppose that catapult in the Punkin Chunkin contest releases a pumpkin 8 feet above the ground at an angle of $45°$ to the horizontal with an initial speed 220 feet per second. The model $s(t) = -16t^2 + 155t + 8$ can be used to estimate the height s of an object after t seconds.

(a) Determine the time at which the pumpkin is at a maximum height.

14a.

(b) Determine the maximum height of the pumpkin.

14b.

(c) After how long will the pumpkin strike the ground?

14c. _____

Five-Minute Warm-Up 7.6
Polynomial Inequalities

In Problems 1 – 4, write in interval notation.

1. $-4 \le x < -2$　　　　　　**2.** $-3 < x \le 1$

1. _____

2. _____

3. $x > -2$　　　　　　**4.** $x \le 5$

3. _____

4. _____

5. Solve: $6x + 3 \le 11x - 7$

5. _____

6. Solve: $(x - 3)(x + 1) = 2$

6. _____

Guided Practice 7.6
Polynomial Inequalities

Objective 1: Solve Quadratic Inequalities

1. A quadratic inequality is an inequality written in one of the following forms:

$ax^2 + bx + c > 0$; $ax^2 + bx + c \geq 0$; $ax^2 + bx + c < 0$; $ax^2 + bx + c \leq 0$.

(a) If $f(x)$ is a quadratic function and $f(x) > 0$, we are interested in finding x values for which the function is _____ (above or below) the x-axis.

1a. _____

(b) If $f(x)$ is a quadratic function and $f(x) < 0$, we are interested in finding x values for which the function is _____ (above or below) the x-axis.

1b. _____

2. We will present two methods for solving quadratic inequalities. These methods are:

2a. _____

2b. _____

3. Solve $x^2 + 2x - 3 \leq 0$ using the graphical method. *(See textbook Example 1)*

Step 1: Write the inequality so that $ax^2 + bx + c$ is on one side of the inequality and 0 is on the other.

The inequality is already in this form: $x^2 + 2x - 3 \leq 0$

Step 2: Graph the function $f(x) = ax^2 + bx + c$. Be sure to label the x-intercepts on the graph.

Function to be graphed: **(a)** _____

x-intercepts: **(b)** _____

Vertex: **(c)** _____

(d) Graph:

Step 3: From the graph, determine where the function is positive and where the function is negative. Use the graph to determine the solution set to the inequality.

Are we looking for positive or negative function values in this problem?   **(e)** _____

State the solution set: **(f)** _____

4a. If $f(x)$ is a quadratic function in factored form and $f(x) > 0$, we are interested in finding the product of the factors for which the function is _____ (positive or negative).

4a. _____

4b. If $f(x)$ is a quadratic function in factored form and $f(x) < 0$, we are interested in finding the product of the factors for which the function is _____ (positive or negative).

4b. _____

5. Solve $x^2 + 2x - 35 > 0$ using the algebraic method. *(See textbook Example 2)*

Step 1: Write the inequality so that $ax^2 + bx + c$ is on one side of the inequality and 0 is on the other.

The inequality is already in this form:

$$x^2 + 2x - 35 > 0$$

Step 2: Determine the solutions to the equation $ax^2 + bx + c = 0$.

Factor: **(a)** _____

Use Zero-Product Property: **(b)** _____

Step 3: Use the solutions to the equation solved in Step 2 to separate the real number line into intervals.

List the intervals: **(c)** _____

Step 4: Write the expression in factored form. Within each interval formed in Step 3, choose a test point and determine the sign of each factor. Then determine the sign of the product. Also determine the value of the expression at each solution found in Step 2.

(d) Complete the chart below:

Interval	$(-\infty,$____$)$	$x =$____	(____,____)	$x =$____	(____, $\infty)$
Test Point					
Sign of ____					
Sign of ____					
Sign of Product					
Conclusion					

To find the solution, select the intervals where the product is (positive or negative)? **(e)** _____

Is 0 included in the solution set? **(f)** _____

State the solution set: **(g)** _____

Do the Math Exercises 7.6
Polynomial Inequalities

In Problems 1 – 10, solve each inequality. Write the solution set in interval notation.

1. $(x-8)(x+1) \leq 0$

2. $(x-4)(x-10) > 0$

1. _____

2. _____

3. $p^2 + 5p + 4 < 0$

4. $2b^2 + 5b < 7$

3. _____

4. _____

5. $x + 6 < x^2$

6. $x^2 - 3x - 5 \geq 0$

5. _____

6. _____

7. $-3p^2 < 3p - 5$

8. $y^2 + 3y + 5 \geq 0$

7. _____

8. _____

9. $3w^2 + w < -2$

10. $p^2 - 8p + 16 \leq 0$

9. _____

10. _____

In Problems 11 and 12, for each function find the values of x that satisfy the given condition.

11. Solve $f(x) > 0$ if $f(x) = x^2 + 4x$

12. Solve $f(x) \leq 0$ if $f(x) = x^2 + 2x - 48$

11. _____

12. _____

In Problems 13 and 14, find the domain of the given function.

13. $f(x) = \sqrt{x^2 - 5x}$

14. $G(x) = \sqrt{x^2 + 2x - 63}$

13. _____

14. _____

15. Revenue Function Suppose that the marketing department of Samsung has found that, when a certain model of cellular telephone is sold a price of p dollars, the daily revenue R (in dollars) as a function of the price p is $R(p) = -5p^2 + 600p$. Determine the prices for which revenue will exceed $17,500. That is, solve $R(p) > 17,500$.

15. _____

In Problems 16 and 17, solve each polynomial inequality.

16. $(3x+4)(x-2)(x-6) \geq 0$

17. $3x^3 + 5x^2 12x - 20 < 0$

16. _____

17. _____

Sullivan/Struve, *Intermediate Algebra*, 3e
Copyright © 2014 Pearson Education, Inc.

Five-Minute Warm-Up 8.1
Composite Functions and Inverse Functions

1. Determine the domain: $R(x) = \dfrac{4}{x^2 - 3x + 2}$.

1. _____

2. Use the function $f(x) = -3x^2 + 1$ to find the following.

(a) $f(-4)$ **(b)** $f(a-2)$

2a. _____

2b. _____

(c) $f(x+h)$

2c. _____

In Problem 3, use the Vertical Line Test to determine whether the following is the graph of a function.

3a.   **3b.**

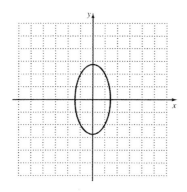   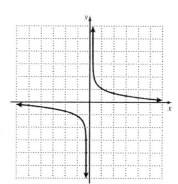

3a. _____

3b. _____

Guided Practice 8.1
Composite Functions and Inverse Functions

Objective 1: Form the Composite Function

1. Function composition is the process of evaluating one function and then carrying the result forward to be evaluated in a second function. The notation $f(g(x))$ means evaluate the inner function, $g(x)$, and then evaluate the result in the outer function, $f(x)$. This function composition is written with the notation

_____. This notation is read "*f* composed with *g*". The mathematical notation "$\circ$" is called small circle, so we can also say "*f* small circle *g* of *x*".

In Problems 2 and 3, suppose that $f(x) = 2x^2 + 7$ *and* $g(x) = x - 1$.
(See textbook Example 1)

2. *Find* $(f \circ g)(-2)$.   **3.** *Find* $(g \circ f)(4)$.   2a. _____

(a) Evaluate $g(-2)$.   **(a)** Evaluate $f(4)$.   2b. _____

                                                                                              2c. _____

(b) Use this result to find $f(g(-2))$.   **(b)** Use this result to find $g(f(4))$.   3a. _____

                                                                                              3b. _____

(c) What is $(f \circ g)(-2)$?   **(c)** What is $(g \circ f)(4)$?   3c. _____

Objective 2: Determine Whether a Function is One-to-One

4. In your own words, what does it mean for a function to be one-to-one?

5. Determine whether or not the function is one-to-one. *(See textbook Example 3)*
    (a) $\{(1, -3), (-3, 1), (2, -4), (-2, 4)\}$ 5a. _____
    (b) $\{(0, 1), (1, 2), (2, 3), (3, 1)\}$ 5b. _____

6. Graphically, how can you test whether a function is one-to-one?   6. _____

Objective 3: Find the Inverse of a Function Defined by a Map or Set of Ordered Pairs

7. In order for the inverse of a function to also be a function, the function must be _____.

8. If $f(x)$ is a one-to-one function, then its inverse function is written with the notation _____.
This means that if the ordered pair (a, b) satisfies $f(x)$, then the ordered pair _____ satisfies the inverse function.

9. Given the function $\{(-3, 15), (-1, -5), (0, 0), (2, 10)\}$, find

(a) the inverse function *(See textbook Example 6)*

9a. _____

(b) the domain of the function

9b. _____

(c) the range of the function

9c. _____

(d) the domain of the inverse function

9d. _____

(e) the range of the inverse function

9e. _____

Objective 4: Obtain the Graph of the Inverse Function from the Graph of the Function

10. Graphs can be transformed in several different ways: translation (slide), reflection (flip), and rotation (turn). A graph is symmetric about a line if one part of the graph is a mirror image of the other. The line on which the graph is flipped (or folded) is called the axis of symmetry. For instance, $y = x^2$ is symmetric about the y-axis. The graph of a function and its inverse are symmetric about the line _____.

Objective 5: Find the Inverse of a Function Defined by an Equation

11. Find the inverse of $f(x) = 4x - 8$. *(See textbook Example 8)*

Step 1: Replace f(x) with y in the equation for f(x).		$f(x) = 4x - 8$
	Replace $f(x)$ with y.	**(a)** _____
Step 2: Interchange the variables to write in inverse.	Rewrite equation (a) exchanging the variables x and y.	**(b)** _____
Step 3: Solve the equation found in Step 2 for y in terms of x.	Add 8 to both sides:	**(c)** _____
	Divide both sides by 4:	**(d)** _____
Step 4: Replace y with $f^{-1}(x)$.		**(e)** _____
Step 5: Verify your result by showing that $f^{-1}(f(x)) = x$ and $f(f^{-1}(x)) = x$.		

Sullivan/Struve, *Intermediate Algebra*, 3e
Copyright © 2014 Pearson Education, Inc.

Name:

Instructor:

Date:

Section:

Do the Math Exercises 8.1
Composite Functions and Inverse Functions

In Problems 1 – 4, use the functions $f(x) = x^2 - 3$ *and* $g(x) = 5x + 1$ *to find each of the following.*

1. $(f \circ g)(3)$　　　　　　　**2.** $(g \circ f)(-2)$

1. _____

2. _____

3. $(f \circ f)(1)$　　　　　　　**4.** $(g \circ g)(-4)$

3. _____

4. _____

5. Given the functions $f(x) = x - 3$ and $g(x) = 4x$, find $(f \circ g)(x)$.

5. _____

6. Given the functions $f(x) = \sqrt{x+2}$ and $g(x) = x - 2$, find $(g \circ f)(x)$.

6. _____

7. Given the functions $f(x) = \dfrac{2}{x-1}$ and $g(x) = \dfrac{4}{x}$, find $(f \circ f)(x)$.

7. _____

In Problems 8 – 10, determine whether the function is one-to-one.

8. $\{(-2, -8), (-1, -1), (0, 0), (1, 1), (2, 8)\}$

8. _____

9. $\{(-3, 0), (-2, 3), (-1, 0), (0, -3)\}$

9. _____

10.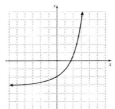

10. _____

11. Find the inverse of the function $\{(-10, 1), (-5, 4), (0, 3), (-5, 2)\}$.

11. _____

In Problems 12 and 13, verify that the functions f and g are inverses of each other.

12. $f(x) = 10x$; $g(x) = \dfrac{x}{10}$

13. $f(x) = \dfrac{2}{x+4}$; $g(x) = \dfrac{2}{x} - 4$

12. _____

13. _____

In Problems 14 – 19, find the inverse function of the given one-to-one function.

14. $g(x) = x + 6$

15. $H(x) = 3x + 8$

14. _____

15. _____

16. $f(x) = x^3 - 2$

17. $G(x) = \dfrac{2}{3-x}$

16. _____

17. _____

18. $R(x) = \dfrac{2x}{x+4}$

19. $g(x) = \sqrt[3]{x+2} - 3$

18. _____

19. _____

20. Volume of a Balloon The volume V of a hot-air balloon (in cubic meters) as a function of its radius r is given by $V(r) = \dfrac{4}{3}\pi r^3$. If the radius r of the balloon is increasing as a function of time t (in minutes) according to $r(t) = 3\sqrt[3]{t}$, for $t \geq 0$, find the volume of the balloon as a function of time t. What will be the volume of the balloon after 30 minutes?

20. _____

Sullivan/Struve, *Intermediate Algebra*, 3e
Copyright © 2014 Pearson Education, Inc.

Five-Minute Warm-Up 8.2
Exponential Functions

In Problems 1 – 4, evaluate each expression.

1. 2^4

2. 2^{-2}

1. _____

2. _____

3. 3^0

4. 10^{-1}

3. _____

4. _____

5. Write 6.023455 as a decimal
(a) rounded to 4 decimal places

(b) truncated to 4 decimal places.

5a. _____

5b. _____

In Problems 6 – 8, simplify each expression.

6. $5x^3 \bullet 2x^{-5}$

7. $\dfrac{4a^2}{10a^{-4}}$

6. _____

7. _____

8. $\left(4p^5\right)^3$

8. _____

9. Solve: $6x^2 = 7x + 5$

9. _____

Guided Practice 8.2
Exponential Functions

Objective 1: Evaluate Exponential Expressions

1. An exponential function is a function of the form $f(x) = a^x$ where a is _____ and $a \neq$ ____.

In Problems 2 and 3, evaluate each expression to 5 decimal places. (See textbook Example 1)

2. $2^{1.4}$ **3.** $2^{\sqrt{2}}$ 2. _____

                                                                3. _____

Objective 2: Graph Exponential Functions

In Problems 4 and 5, match the function to the graph.
(See textbook Examples 2 and 3)

4. $f(x) = a^x;\, a > 1$ **(a)**   4. _____

5. $f(x) = a^x;\, 0 < a < 1$ 5. _____

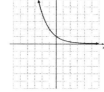

(b)

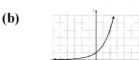

Objective 3: Define the Number e

                                                                6. _____

6. What is an approximate value of e to the nearest thousandth?

7. Evaluate each of the following to two decimal places.

(a) e^3 **(b)** e^{-2} 7a. _____

                                                                7b. _____

Objective 4: Solve Exponential Equations

8. The Property for Solving Exponential Equations states that if two exponential functions have the same

base and the exponential functions are equal, then it must be true that _____.

9. Solve: $3^{2x+1} = 27$ *(See textbook Example 5)*

Step 1: Use the Laws of Exponents to write both sides of the equation with the same base.

Prime factor 27 and write in exponential form:

$3^{2x+1} = 27$

(a) _____

Step 2: Set the exponents on each side of the equation equal to each other.

(b) _____

Step 3: Solve the equation resulting from Step 2.

Subtract 1 from both sides:

(c) _____

Divide both sides by 2:

(d) _____

Step 4: Verify your solution(s). Substitute your value into the original equation.

We leave it to you to verify your solution.

State the solution set:

(e) _____

Objective 5: Use Exponential Models that Describe Our World

10. Strontium 90 is a radioactive material that decays according to the function $A(t) = A_0 e^{-0.087t}$, where A_0 is the initial amount present and A is the amount present at time t (in days). If you begin with 100 grams of Strontium 90, how much will be present after 5 days?

10. _____

11. What is the compound interest formula? Identify each of the variables in the formula.

12. Suppose that you deposit $100 into a savings account that earns 7% compounded quarterly for a period of 5 years.

(a) Identify the value of each variable in the compound interest formula.

12a. _____

(b) How much is in the account 5 years after the $100 deposit?

12b. _____

(c) You also deposit $100 in another bank because you believe you are getting a better deal when you receive 7% compounded daily. Assuming there are 360 days per business year, identify the value of each variable in the compound interest formula.

12c. _____

(d) How much is in the account after 5 years?

12d. _____

Sullivan/Struve, *Intermediate Algebra*, 3e
Copyright © 2014 Pearson Education, Inc.

Do the Math Exercises 8.2
Exponential Functions

In Problem 1 and 2, approximate each number using a calculator. Round your answer to three decimal places.

1a. $5^{1.4}$

1b. $5^{1.41}$

1a. _____

1b. _____

1c. $5^{1.414}$

1d. $5^{1.4142}$

1c. _____

1d. _____

1e. $5^{\sqrt{2}}$

1e. _____

2a. e^3

2b. $e^{1.5}$

2a. _____

2b. _____

In Problems 3 – 5, graph each function.

3. $g(x) = 10^x$

4. $H(x) = 2^{x-2}$

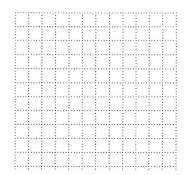

5. $f(x) = e^x - 1$

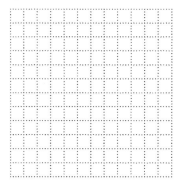

In Problems 6–11, solve each equation.

6. $3^x = 3^{-2}$

7. $2^{x+3} = 128$

6. _____

7. _____

8. $3^{-x+4} = 27^x$

9. $9^{2x} \cdot 27^{x^2} = 3^{-1}$

8. _____

9. _____

10. $\left(\dfrac{1}{6}\right)^x - 36 = 0$

11. $e^{3x} = e^2$

10. _____

11. _____

In Problem 12 and 13, suppose that $f(x) = 3^x$.

12. What is $f(2)$? What point is on the graph of f?

13. If $f(x) = \dfrac{1}{81}$, what is x? What point is on the graph of f?

12. _____

13. _____

14. A Population Model According to the U.S. Census Bureau, the population of the world in 2012 was 7018 million people. In addition, the population of the world was growing at a rate of 1.26% per year. Assuming that this growth rate continues, the model $P(t) = 7018(1.0126)^{t-2012}$ represents the population P (in millions of people) in year t.

(a) According to this model, what will be the population of the world in 2015?

14a. _____

(b) According to this model, what will be the population of the world in 2025?

14b. _____

Five-Minute Warm-Up 8.3
Logarithmic Functions

1. Solve: $4x + 6 > 0$ 1. _____

2. Solve: $\sqrt{2x + 3} = x$ 2. _____

3. Solve: $x^2 = -7x - 12$ 3. _____

In Problems 4 and 5, evaluate each expression.

4. 4^{-2} **5.** $\left(\dfrac{1}{2}\right)^{-3}$ 4. _____

                                                        5. _____

In Problems 6 and 7, find the function value.

6. $f(x) = x^2;\ f(-2)$ **7.** $f(x) = 2^x;\ f(-2)$ 6. _____

                                                        7. _____

Guided Practice 8.3
Logarithmic Functions

1. The logarithmic function to the base a, where $a > 0$ and $a \neq 1$, is denoted by $y = \log_a x$ and is read as "y is the logarithm to the base a of x". This is defined as

$$y = \log_a x \text{ is equivalent to } \underline{\hspace{4cm}}$$

Objective 1: Change Exponential Equations to Logarithmic Equations

In Problems 2 and 3, rewrite each exponential expression as an equivalent expression involving a logarithm. (See textbook Example 1)

2. $2^{-3} = \dfrac{1}{8}$ 　　　　　　　　　**3.** $a^4 = 3$

2. _____

3. _____

Objective 2: Change Logarithmic Equations to Exponential Equations

In Problems 4 and 5, rewrite each logarithmic expression as an equivalent expression involving an exponent. (See textbook Example 2)

4. $p = \log_4 30$ 　　　　　　　　　**5.** $\log_n 3 = -5$

4. _____

5. _____

Objective 4: Determine the Domain of a Logarithmic Function
(See textbook Example 5)

6a. The domain of the logarithmic function is

6b. The range of the logarithmic function is

6c. Determine the domain of the function $f(x) = \log_3(10 - 2x)$.

6a. _____

6b. _____

6c. _____

Objective 5: Graph Logarithmic Functions

7. Because exponential functions and logarithmic functions are inverses of each other, we know that the

graph of $y = \log_5 x$ is a reflection of _____ across the line $y = x$. Graph both of these functions in the same coordinate plane. *(See textbook Example 6)*

Objective 6: Work with Natural and Common Logarithms

8. The natural logarithm: $y = \ln x$ is written in exponential form as

8. _____

9. The common logarithm: $y = \log x$ is written in exponential form as:

9. _____

Objective 7: Solve Logarithmic Equations

In Problems 10 and 11, solve the logarithmic equation. Be sure to check your answers as extraneous solutions may appear. (See textbook Examples 9 and 10)

10. $\log_3 (5 - 2x) = 2$ **11.** $\ln x = -2$

10. _____

11. _____

Objective 8: Use Logarithmic Models That Describe Our World

12. List 3 applications of logarithmic functions that are used in the world today.

12a. _____

12b. _____

12c. _____

Sullivan/Struve, *Intermediate Algebra*, 3e
Copyright © 2014 Pearson Education, Inc.

Do the Math Exercises 8.3
Logarithmic Functions

In Problems 1 – 3, change each exponential expression to an equivalent expression involving a logarithm.

1. $64 = 4^3$

2. $b^4 = 23$

1. _____

2. _____

3. $10^{-3} = z$

3. _____

In Problems 4 – 6, change each logarithmic expression to an equivalent expression involving an exponent.

4. $\log_3 81 = 4$

5. $\log_6 x = -4$

4. _____

5. _____

6. $\log_a 16 = 2$

6. _____

In Problems 7 – 8, find the exact value of each logarithm without using a calculator.

7. $\log_4 16$

8. $\log_{\sqrt{3}} 3$

7. _____

8. _____

In Problems 9 – 10, find the domain of each function.

9. $f(x) = \log_3(x - 2)$

10. $G(x) = \log_4(3 - 5x)$

9. _____

10. _____

In Problems 11 and 12, use a calculator to evaluate each expression. Round your answer to three decimal places.

11. $\log 0.78$

12. $\ln \dfrac{1}{2}$

11. _____

12. _____

In Problems 13 – 18, solve each logarithmic equation.

13. $\log_3(5x - 3) = 3$

14. $\log_4(8x + 10) = 3$

13. _____

14. _____

15. $\log_a 81 = 2$

16. $\log (2x + 3) = 1$

15. _____

16. _____

17. $\ln e^{2x} = 8$

18. $\log_3(x^2 + 1) = 2$

17. _____

18. _____

19. Alaska, 1964 According to the United States Geological Survey, an earthquake on March 28, 1964 in Prince William Sound, Alaska resulted in a seismographic reading of 1,584,893 millimeters 100 kilometers from its epicenter. What was the magnitude of this earthquake? This earthquake was the second largest ever recorded, with the largest being the Great Chilean Earthquake of 1960, whose magnitude was 9.5 on the Richter scale.

19. _____

Five-Minute Warm-Up 8.4
Properties of Logarithms

1. Write 1.13985 as a decimal

(a) rounded to 3 decimals places   **(b)** truncated to 3 decimal places.   1a. _____

1b. _____

In Problems 2 and 3, write each expression with a rational exponent.

2. $\sqrt{x}$ **3.** $\sqrt[4]{a^3}$ 2. _____

3. _____

In Problems 4 and 5, write in exponential form using prime numbers in the base.

4. $\dfrac{1}{16}$ **5.** $\dfrac{27}{8}$ 4. _____

5. _____

In Problems 6 and 7, simplify each expression.

6. $x^0, x \neq 0$ **7.** $6a^0, a \neq 0$ 6. _____

7. _____

8. Find the inverse function of $y = \log_3 x$. Write the equation in exponential form.   8. _____

Guided Practice 8.4
Properties of Logarithms

Objective 1: Understand the Properties of Logarithms

1. There are four useful properties that can be quickly derived from the definition of a logarithm. While it is not essential, it will make your work with logarithms faster and easier if you memorize them.

For any real number a, for which the logarithm is defined:

$$\log_a 1 = 0 \qquad \log_a a = 1 \qquad \log_a a^r = r \qquad a^{\log_a M} = M$$

Use these properties to simplify the following. *(See textbook Examples 1 and 2)*

(a) $3^{\log_3 \sqrt{2}}$ **(b)** $\log_4 1$ **(c)** $\ln e$ **(d)** $\log_x x^7$

1a. _____

1b. _____

1c. _____

1d. _____

Objective 2: Write a Logarithmic Expression as a Sum or Difference of Logarithms

2. There are three important rules that you will need as well. These should be easy to remember because they are very similar to rules you already know, Laws of Exponents. For each of these, M, N, and a are positive real numbers with $a \neq 1$.

(a) The Product Rule: $\log_a (M \cdot N) =$ _____

(b) The Quotient Rule: $\log_a \left(\dfrac{M}{N} \right) =$ _____

(c) The Power Rule: $\log_a M^r =$ _____

3. Use the Product Rule of Logarithms to write $\log(9x)$ as the sum of logarithms. *(See textbook Example 4)*

3. _____

4. Use the Quotient Rule of Logarithms to write $\ln\left(\dfrac{2}{x}\right)$ as the difference of logarithms.

(See textbook Example 5)

4. _____

Objective 3: Write a Logarithmic Expression as a Single Logarithm

5. Now we will reverse the process and write an expanded logarithmic expression as a single logarithm. This will be an important skill in the next section.

Use the Quotient Rule of Logarithms to write $\log_4 (x + 1) - \log_4 (x^2 - 1)$ as a single logarithm.
(See textbook Example 9)

5. _____

Objective 4: Evaluate Logarithms Whose Base is Neither 10 Nor e

6. Calculators only have keys for the common logarithm, log, and the natural logarithm, ln. When we need to find the logarithmic value of an expression that uses a base other than 10 or e, we use the Change-of-Base Formula. If M, a and b are positive real numbers with $a \neq 1$, $b \neq 1$, then

Change-of-Base Formula: $\log_a M =$

_____

7. Approximate $\log_2 9$ using the Change-of-Base Formula. Round you answer to three decimal places.
(See textbook Example 12)

7. _____

8. Write an examples to illustrate: $\log_2 (x + y) \neq \log_2 x + \log_2 y$

Sullivan/Struve, *Intermediate Algebra*, 3e
Copyright © 2014 Pearson Education, Inc.

Do the Math Exercises 8.4
Properties of Logarithms

In Problem 1, use properties of logarithms to find the exact value of each expression. Do not use a calculator.

1a. $\log_5 5^{-3}$

1b. $5^{\log_5 \sqrt{2}}$

1c. $e^{\ln 10}$

1a. _____

1b. _____

1c. _____

In Problem 2, suppose that $\ln 2 = a$ *and* $\ln 3 = b$. *Use properties of logarithms to write each logarithm in terms of a and b.*

2a. $\ln 4$

2b. $\ln 18$

2a. _____

2b. _____

In Problems 3 – 10, write each expression as a sum and/or difference of logarithms. Express exponents as factors.

3. $\log_4 \left(\dfrac{a}{b} \right)$

4. $\log_3 (a^3 b)$

3. _____

4. _____

5. $\log_2 (8z)$

6. $\log_2 \left(\dfrac{16}{p} \right)$

5. _____

6. _____

7. $\log_2 \left(32 \sqrt[4]{z} \right)$

8. $\ln \left(\dfrac{\sqrt[5]{x}}{(x+2)^2} \right)$

7. _____

8. _____

9. $\log_6 \sqrt[3]{\dfrac{x-2}{x+1}}$

10. $\log_4\left[\dfrac{x^3(x-3)}{\sqrt[3]{x+1}}\right]$

9. _____

10. _____

In Problems 11 – 19, write each expression as a single logarithm.

11. $\log_4 32 + \log_4 2$

12. $\log_2 48 - \log_2 3$

11. _____

12. _____

13. $8\log_2 z$

14. $4\log_2 a + 2\log_2 b$

13. _____

14. _____

15. $\dfrac{1}{3}\log_4 z + 2\log_4(2z+1)$

16. $\log_7 x^4 - 2\log_7 x$

15. _____

16. _____

17. $\dfrac{1}{3}\left[\ln(x-1)+\ln(x+1)\right]$

18. $\log_5\left(x^2+3x+2\right)-\log_5\left(x+2\right)$

17. _____

18. _____

19. $10\log_4 \sqrt[5]{x} + 4\log_4 \sqrt{x} - \log_4 16$

19. _____

In Problem 20, use the Change-of-Base Formula and a calculator to evaluate each logarithm. Round your answer to three decimal places.

20a. $\log_7 5$

20b. $\log_{\sqrt{3}} \sqrt{6}$

20a. _____

20b. _____

Sullivan/Struve, *Intermediate Algebra*, 3e
Copyright © 2014 Pearson Education, Inc.

Five-Minute Warm-Up 8.5
Exponential and Logarithmic Equations

In Problems 1 – 4, solve each equation.

1. $\dfrac{2}{3}x - 7 = -1$

2. $x^2 + 5x = 24$

1. _____

2. _____

3. $4n^2 = 2 - 7n$

4. $(x-1)^2 - 5(x-1) - 6 = 0$

3. _____

4. _____

5. Find the domain: $f(x) = \log_2(-2x + 6)$

5. _____

Guided Practice 8.5
Exponential and Logarithmic Equations

Objective 1: Solve Logarithmic Equations Using the Properties of Logarithms

1. We use the following property where *M*, *N*, and *a* are positive real numbers with $a \neq 1$ to solve logarithmic equations where the log function appears on both sides of the equation.

$$\text{If } \log_a M = \log_a N \text{, then } M = N.$$

M and *N* are called arguments so we say, "if there is equality between two logarithmic expressions which have the same base, set the arguments equal." Note that each log function has to be completely simplified to a single logarithm. Also, be careful to check for extraneous solutions.

Use this property to solve the following equations. *(See textbook Examples 1 and 2)*

(a) $\log_2 x = \log_2 (3x - 5)$

(b) $\log_4 (x + 3) - \log_4 x = \log_4 10$

1a. _____

1b. _____

(c) $\dfrac{1}{2}\ln x = 3\ln 2$

1c. _____

Objective 2: Solve Exponential Equations

2. In Section 8.2, we were able to solve exponential equations of the form $a^u = a^v$ by using the Property for Solving Exponential Equations. This states that if two exponential functions have the same base and the exponential functions are equal, it must be true that the exponents are equal.

Now we will encounter exponential equations which cannot be written on a common base. For this circumstance, we use the definition of a logarithm to convert from exponential form to logarithmic form. Use this approach to solve each of the following. Give both the exact and approximate solution. *(See textbook Examples 3 and 4)*

(a) $3^x = 12$

(b) $\dfrac{1}{2}e^{3x} = 9$

2a. _____

2b. _____

Objective 3: Solve Equations Involving Exponential Models

3. Radioactive Decay The half-life of carbon-10 is 19.255 seconds. Suppose that a researcher possesses a 200-gram sample of carbon-10. The amount A (in grams) of carbon-10 after t seconds is given by

$$A(t) = 100 \cdot \left(\frac{1}{2}\right)^{\frac{t}{19.255}}$$

(See textbook Example 5)

(a) Write a model to find when there will be 90 grams of carbon-10 left in the sample.   3a. _____

(b) Use logarithms to solve the equation from part (a).   3b. _____

4. The population of a small town is growing at a rate of 3% per year. The population of the town can be calculated by the exponential function $P(t) = 2500e^{0.03t}$ where t is the number of years after 1950. *(See textbook Examples 5 and 6)*

(a) What was the population in 1965?   4a. _____

(b) Write a model to find when the population reached 7500 people.   4b. _____

(c) Use logarithms to solve the equation from (b) and find when the population reached 7500 people.   4c. _____

(d) How long will it take for the population to double? (That is, for the town to have a population of 5000 people.)   4d. _____

Sullivan/Struve, *Intermediate Algebra*, 3e
Copyright © 2014 Pearson Education, Inc.

Do the Math Exercises 8.5
Exponential and Logarithmic Equations

In Problems 1 – 18, solve each equation. Express irrational solutions in exact form and as a decimal rounded to 3 decimal places.

1. $\log_5 x = \log_5 13$

2. $\dfrac{1}{2}\log_2 x = 2\log_2 2$

1. _____

2. _____

3. $\log_2 (x - 7) + \log_2 x = 3$

4. $\log_3 (x + 5) - \log_3 x = 2$

3. _____

4. _____

5. $\log_5 (x + 3) + \log_5 (x - 4) = \log_5 8$

6. $3^x = 8$

5. _____

6. _____

7. $4^x = 20$

8. $e^x = 3$

7. _____

8. _____

9. $10^x = 0.2$

10. $2^{2x} = 5$

9. _____

10. _____

11. $3 \cdot 4^x = 15$

12. $\log_6 x + \log_6 (x + 5) = 2$

11. _____

12. _____

13. $3 \log_2 x = \log_2 8$

14. $5 \log_4 x = \log_4 32$

13. _____

14. _____

15. $-4e^x = -16$

16. $9^x = 27^{x-4}$

15. _____

16. _____

17. $\log_7 x^2 = \log_7 8$

18. $\log_3 (x - 5) + \log_3 (x + 1) = \log_3 7$

17. _____

18. _____

19. A Population Model According to the *United States Census Bureau*, the population of the world in 2012 was 7018 million people. In addition, the population of the world was growing at a rate of 1.26% per year. Assuming that this growth rate continues, the model $P(t) = 7018(1.0126)^{t - 2012}$ represents the population P (in millions of people) in year t. According to this model, when will the population of the world be 11.58 billion people?

19. _____

20. Depreciation Based on data obtained from the *Kelley Blue Book*, the value V of a Chevy Malibu that is t years old can be modeled by $V(t) = 25{,}258(0.84)^t$. According to the model, when will the car be worth $15,000?

20. _____

Five-Minute Warm-Up 9.1
Distance and Midpoint Formulas

In Problems 1 – 3, simplify each expression.

1. $\sqrt{40}$　　　　　　**2.** $\sqrt{108}$　　　　　　**3.** $\sqrt{32p^4}$

1.＿＿＿＿＿＿

2.＿＿＿＿＿＿

3.＿＿＿＿＿＿

4. Simplify the expression: $\sqrt{(2x-5)^2}$

4.＿＿＿＿＿＿

In Problems 5 and 6 simplify each expression.

5. $-2\sqrt{25}$　　　　　　　　　**6.** $\sqrt{(-5-2)^2 + (16-(-8))^2}$

5.＿＿＿＿＿＿

6.＿＿＿＿＿＿

7. Find the area of a triangle whose base has length of $\sqrt{18}$ cm and whose height has length of $\sqrt{8}$ cm.

7.＿＿＿＿＿＿

8. Find the length of the hypotenuse of a right triangle whose legs are 6 and $6\sqrt{3}$.

8.＿＿＿＿＿＿

Guided Practice 9.1
Distance and Midpoint Formulas

● *Objective 1: Use the Distance Formula*

1. We can find the distance between two points in the Cartesian plane using the Pythagorean Theorem. This can be accomplished by plotting the points, drawing a right triangle, and applying $a^2 + b^2 = c^2$ where c is the length of the hypotenuse or, in this case, the distance between the points. If you get stuck or forget the distance formula, you can always use this approach.

We use the Distance Formula to find the length of a line segment quickly and easily. If the two points in the Cartesian plane are $P_1 = (x_1, y_1)$ and $P_2 = (x_2, y_2)$, the distance between P_1 and P_2, denoted $d(P_1, P_2)$, is

_____

2. Be sure to use the Rule for Order of Operations when finding the distance between two points. If we want to find the distance between (a, b) and (c, d), the steps are:

(a) subtract _____; **(b)** subtract _____; **(c)** square the value from part _____;

● **(d)** square the value from part _____; **(e)** _____; **(f)** _____; **(g)** _____.

3. Find the distance between $(9, 3)$ and $(1, -1)$. Find both the exact value and the approximate distance to two decimal places.   *(See textbook Example 1)*

3. _____

●

Objective 2: Use the Midpoint Formula

4. A *midpoint* is a point (in this case an ordered pair) which divides a line segment into _____.

5. To find the coordinates of the midpoint, we use the Midpoint Formula. This states that if a line segment has endpoints at $P_1 = (x_1, y_1)$ and $P_2 = (x_2, y_2)$, the midpoint, M, is an ordered pair such that

$$M = \left(\frac{}{2}, \frac{}{2} \right).$$

You can see that the Midpoint Formula averages the x values to find a coordinate in the middle and averages the y values to find a coordinate in the middle. You can verify that the midpoint divides the segment into two segments of equal length by finding the distance from P_1 to M and the finding the distance from M to P_2. These distances will be equal. This step is not necessary, but it is a good check.

6. Find the midpoint of the line segment joining $P_1 = (-3, 2)$ and $P_2 (-5, -8)$. *(See textbook Example 4)*

6. _____

Sullivan/Struve, *Intermediate Algebra*, 3e
Copyright © 2014 Pearson Education, Inc.

Do the Math Exercises 9.1
Distance and Midpoint Formulas

In Problems 1 – 6, find the distance $d(P_1, P_2)$ between points P_1 and P_2.

1. $P_1 = (1, 3)$; $P_2 = (4, 7)$ **2.** $P_1 = (-10, -3)$; $P_2 = (14, 4)$

1. _____

2. _____

3. $P_1 = (-1, 2)$; $P_2 = (-1, 0)$ **4.** $P_1 = (5, 0)$; $P_2 = (-1, -4)$

3. _____

4. _____

5. $P_1 = \left(\sqrt{6}, -2\sqrt{2}\right)$; $P_2 = \left(3\sqrt{6}, 10\sqrt{2}\right)$ **6.** $P_1 = (-1.7, 1.3)$; $P_2 = (0.3, 2.6)$

5. _____

6. _____

In Problems 7 – 12, find the midpoint of the line segment formed by joining points P_1 and P_2.

7. $P_1 = (1, 3)$; $P_2 = (5, 7)$ **8.** $P_1 = (-10, -3)$; $P_2 = (14, 7)$

7. _____

8. _____

9. $P_1 = (-1, 2)$; $P_2 = (3, 9)$ **10.** $P_1 = (5, 0)$; $P_2 = (-1, -4)$

9. _____

10. _____

11. $P_1 = \left(\sqrt{6}, -2\sqrt{2}\right)$; $P_2 = \left(3\sqrt{6}, 10\sqrt{2}\right)$ **12.** $P_1 = (-1.7, 1.3)$; $P_2 = (0.3, 2.6)$

11. _____

12. _____

13. Consider the three points $A = (-2, 3)$, $B = (2, 0)$, and $C = (5, 6)$.

(a) Plot each point in the Cartesian plane and form the triangle ABC.

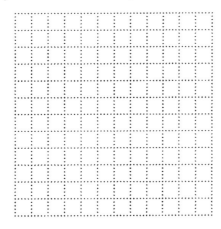

13b. _____

(b) Find the length of each side of the triangle.

14. Find all points having a y-coordinate of -3 whose distance from the point $(-4, 2)$ is 13.　　14. _____

Five-Minute Warm-Up 9.2
Circles

In Problems 1 and 2, complete the square. Determine the number that must be added to the expression to make it a perfect square trinomial. Then factor the expression.

1. $x^2 - 14x$ **2.** $y^2 - 5y$

1. _____

2. _____

In Problems 3 and 4, factor completely.

3. $y^2 + 16y + 64$ **4.** $2x^2 - 12x + 18$

3. _____

4. _____

5. Find **(a)** the area and **(b)** the circumference of a circle whose diameter is 15 inches. Give both the exact answer and then the answer rounded to 2 decimal places.

5a. _____
_____

5b. _____
_____

Name: Date:
Instructor: Section:

Guided Practice 9.2
Circles

Objective 1: Write the Standard Form of the Equation of a Circle

1. A *circle* is the set of all points in the Cartesian plane that are a fixed distance r from a fixed point (h, k).

(a) The point (h, k) is called the _____.

(b) The fixed distance r is called the _____.

(c) We also know that if d is the length of the diameter of the circle, then _____.

2. The *standard form of an equation of a circle* with radius r and center (h, k) is _____.

3. Write the standard form of the equation of the circle with radius 5 and center $(-1, 3)$.
(See textbook Example 1)

3. _____ _____

Objective 2: Graph a Circle

4. Graph the equation: $(x - 2)^2 + (x + 3)^2 = 4$ *(See textbook Example 2)*

(a) Identify the center: _____ (b) Length of radius: _____ (c) Graph:

Objective 3: *Find the Center and Radius of a Circle Given an Equation in General Form*

5. General form expands (multiplies) the binomials, regroups like terms, and has all the terms on one side of the equation and zero on the other side of the equation. What is the general form of the equation of a circle?

_____

6. Graph the equation: $x^2 + y^2 + 6x - 2y + 1 = 0$ *(See textbook Example 3)*

(a) Complete the square for the *x* terms and complete the square for the *y* terms:

$$x^2 + 6x + \underline{\quad} + y^2 - 2y + \underline{\quad} = -1 + \underline{\quad} + \underline{\quad}$$

(b) Factor:

$$(x + \underline{\quad})^2 + (y - \underline{\quad})^2 = \underline{\quad}$$

(c) Identify the center: _____ and the length of radius: _____

(d) Graph:

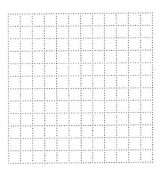

7. Are circles functions? Why or why not?

8. Is $3x^2 - 12x + 3y^2 - 15 = 0$ The equation of a circle? Why or why not? If so, what is the center and radius?

8. _____

Do the Math Exercises 9.2
Circles

In Problems 1 – 4, write the standard form of the equation of each circle whose radius is r and center is (h, k). Graph each circle.

1. $r = 5; (h, k) = (0, 0)$

2. $r = 2; (h, k) = (1, 0)$

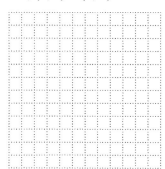

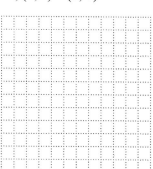

1. _____

2. _____

3. $r = 4; (h, k) = (-4, 4)$

4. $r = \sqrt{7}; (h, k) = (5, 2)$

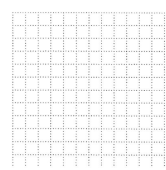

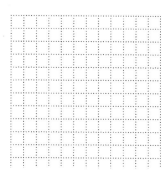

3. _____

4. _____

In Problems 5 – 8, identify the center (h, k) and radius r of each circle. Graph each circle.

5. $x^2 + y^2 = 25$

6. $(x - 5)^2 + (y + 2)^2 = 49$

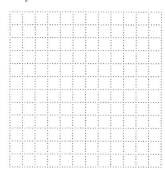

5. _____

6. _____

7. $(x - 6)^2 + y^2 = 36$

8. $(x - 2)^2 + (y + 2)^2 = \dfrac{1}{4}$

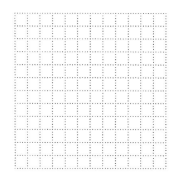

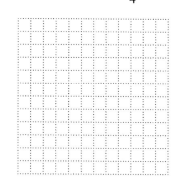

7. _____

8. _____

In Problems 9 – 11, find the center (h, k) and radius r of each circle.

9. $x^2 + y^2 + 2x - 8y + 8 = 0$

10. $x^2 + y^2 + 4x - 12y + 36 = 0$

9. _____

10. _____

11. $2x^2 + 2y^2 - 28x + 20y + 20 = 0$

11. _____

In Problems 12 – 14, find the standard form of the equation of each circle.

12. Center at (0, 3) and containing the point (3, 7).

12. _____

13. Center at (2, –3) and tangent to the *x*-axis.

13. _____

14. With endpoints of a diameter at (–5, –3) and (7, 2).

14. _____

15. Find the area and circumference of the circle $(x - 1)^2 + (y - 4)^2 = 49$.

15. _____

Five-Minute Warm-Up 9.3
Parabolas

In Problems 1 – 3, use the function $f(x) = 2(x-2)^2 + 1$.

1. Identify the vertex.

2. Does the parabola open up or down?

1. _____

2. _____

3. Name the axis of symmetry.

3. _____

In Problems 4 and 5, complete the square. Determine the number that must be added to the expression to make it a perfect square trinomial. Then factor the expression.

4. $x^2 + 10x$

5. $-3x^2 + 12x$

4. _____

5. _____

6. Solve: $(x-4)^2 = 16$

6. _____

Guided Practice 9.3
Parabolas

1. A *parabola* is the set of all points P in the plane that are the same distance from a fixed point F as they are from a fixed line D. In other words, a parabola is the set points P such that $d(F, P) = d(P, D)$.

    (a) The point F is called the _____ of the parabola.

    (b) The line D is its _____.

    (c) The turning point of the parabola is its _____.

    (d) The line through the point F and perpendicular to the line D is called the _____.

Objective 1: Graph Parabolas Whose Vertex Is the Origin

2. In this course, the axis of symmetry is parallel to either the x-axis or y-axis. This means that the parabola opens up, down, left, or right. In the equation of the parabola, the coefficient of the linear variable $\left(x^1 \text{ or } y^1\right)$ will determine the direction that the parabola opens. If k is a real number (in the text $k = 4a$ as this coefficient can be used to determine the breadth of the parabola):

(a) $y^2 = kx$ opens either

    2a. _____

    (b) If $k > 0$, the parabola opens

    2b. _____

    (c) If $k < 0$, the parabola opens

    2c. _____

Notice that in (b), the positive x-axis has an arrow which points to the right and the parabola opens right.

(d) $x^2 = ky$ opens either

    2d. _____

    (e) If $k > 0$, the parabola opens

    2e. _____

    (f) If $k < 0$, the parabola opens

    2f. _____

Notice that in (e), the positive y-axis has an arrow which points up and the parabola opens up.

3. Determine which direction each parabola opens. *(See textbook Examples 1 and 2)*

(a) $x^2 = -4y$ **(b)** $y^2 = 4x$

    3a. _____

    3b. _____

Objective 2: Find the Equation of a Parabola

4. Find the equation of a parabola with vertex at $(0, 0)$ if its axis of symmetry is the *x*-axis and its graph contains the point $(-2, -1)$. *(See textbook Examples 3 and 4)*

(a) Which direction does the parabola open?

4a. _____

(b) Review the equations on page 702 of your text. Which matches the given information?

4b. _____

4c. _____

(c) Substitute the given information to find the equation for the parabola.

Objective 3: Graph a Parabola Whose Vertex Is Not the Origin

5. Graph the parabola $x^2 - 8x + 4y + 20 = 0$. *(See textbook Example 5)*

(a) Isolate the terms involving the second-degree variable:

5a. _____

(b) Complete the square:

5b. _____

(c) Simplify:

5c. _____

(d) Factor:

5d. _____

(e) Which direction does the parabola open?

5e. _____

(f) Identify the vertex.

5f. _____

(g) Find two more points on the parabola and then graph.

Sullivan/Struve, *Intermediate Algebra*, 3e
Copyright © 2014 Pearson Education, Inc.

Do the Math Exercises 9.3
Parabolas

In Problems 1 – 5, find the equation of the parabola described.

1. vertex at $(0, 0)$; focus at $(0, 5)$

1. _____

2. vertex at $(0, 0)$; focus at $(-8, 0)$

2. _____

3. vertex at $(0, 0)$; contains the point $(2, 2)$; axis of symmetry the x-axis

3. _____

4. vertex at $(0, 0)$; directrix $x = -4$

4. _____

5. focus at $(0, -2)$; directrix $y = 2$

5. _____

In Problems 6 – 13, graph the parabola. Find (a) the vertex, (b) the focus, and (c) the directrix.

6. $x^2 = 28y$ **7.** $y^2 = 10x$

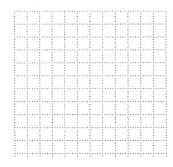   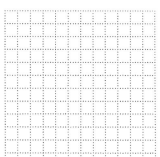

6a. _____

6b. _____

6c. _____

7a. _____

7b. _____

7c. _____

8. $x^2 = -16y$

9. $(x + 4)^2 = -4(y - 1)$

8a. _____

8b. _____

8c. _____

9a. _____

9b. _____

9c. _____

10. $(y - 2)^2 = 12(x + 5)$

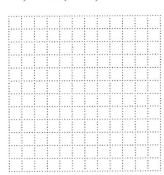

11. $x^2 + 2x - 8y + 25 = 0$

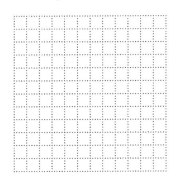

10a. _____

10b. _____

10c. _____

11a. _____

11b. _____

11c. _____

12. $y^2 - 8y + 16x - 16 = 0$

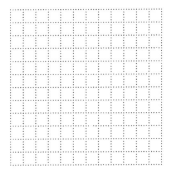

13. $x^2 - 4x + 10y + 4 = 0$

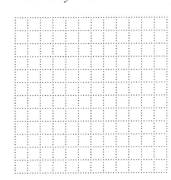

12a. _____

12b. _____

12c. _____

13a. _____

13b. _____

13c. _____

14. Suspension Bridge The cables of a suspension bridge are in the shape of a parabola. The towers supporting the cable are 400 feet apart and 80 feet high. If the cables touch the road surface midway between the towers, what is the height of the cable at a point 100 feet from the center of the bridge?

14. _____

Sullivan/Struve, *Intermediate Algebra*, 3e
Copyright © 2014 Pearson Education, Inc.

Five-Minute Warm-Up 9.4
Ellipses

In Problems 1 – 3, complete the square. Determine the number that must be added to the expression to make it a perfect square trinomial. Then factor the expression.

1. $y^2 - 9y$

2. $x^2 + 12x$

1. _____

2. _____

3. $4y^2 - 2y$

3. _____

In Problems 4 – 7, use the function $y = 9x^2 + 54x - 87$.

4. Write the equation in standard form by completing the square.

5. Identify the vertex.

4. _____

5. _____

6. _____

6. Does the parabola open up or down?

7. Name the axis of symmetry.

7. _____

Guided Practice 9.4
Ellipses

Objective 1: Graph an Ellipse Whose Center Is the Origin

1. An *ellipse* is the set of all points in the plane such that the sum of the distances from two fixed points is a constant.

   (a) The fixed points are called the _____.

   (b) The long axis contains the fixed points and is called the _____.

   (c) The other axis is perpendicular to long axis and is called the _____.

   (d) The point where the two axes intersect is the _____ of the ellipse.

   (e) The major axis contains turning points of the ellipse called _____.

2. The standard form of an ellipse is either $\dfrac{x^2}{a^2} + \dfrac{y^2}{b^2} = 1$ or $\dfrac{x^2}{b^2} + \dfrac{y^2}{a^2} = 1$.
(See textbook Examples 1 and 2)

   (a) In either of these equations, the larger of the two denominators is _____.

   (b) The distance from the center to the vertex is _____ units of length.

   (c) Therefore, the length of the major axis is _____.

   (d) The length of the minor axis is _____.

   (e) The term with larger denominator tell us which axis is the _____ axis.

   (f) The foci are c units from the center, on the major axis, where $c^2 = $ _____.

   (g) For these two equations, the center of the ellipse is _____.

3. Find the intercepts to graph each ellipse: *(See textbook Examples 1 and 2)*

(a) $\dfrac{x^2}{9} + \dfrac{y^2}{25} = 1$ **(b)** $4x^2 + 16y^2 = 64$

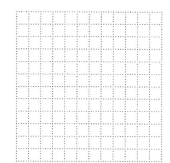   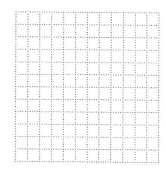

Objective 2: Find the Equation of an Ellipse Whose Center Is the Origin

4. Find the equation, in standard form, of the ellipse whose center is at $(0, 0)$, one focus is at $(0, 5)$ and one vertex is at $(0, 13)$. *(See textbook Example 3)*

4. _____

Objective 3: Graph an Ellipse Whose Center Is Not the Origin

In Problems 5 – 7, **(a)** write the equation in the form $\dfrac{(x-h)^2}{a^2} + \dfrac{(y-k)^2}{b^2} = 1$ or

$\dfrac{(x-h)^2}{b^2} + \dfrac{(y-k)^2}{a^2} = 1$, then identify **(b)** the center, **(c)** the vertices and **(d)** the foci.

(See textbook Example 4)

5a. _____

5b. _____

5c. _____

5. $\dfrac{(x+5)^2}{9} + \dfrac{(y-2)^2}{25} = 1$

5d. _____

6a. _____

6b. _____

6. $4(x-1)^2 + 9(y-3)^2 = 36$

6c. _____

6d. _____

7. $16x^2 + 9y^2 - 128x + 54y - 239 = 0$

7a. _____

7b. _____

7c. _____

7d. _____

Do the Math Exercises 9.4
Ellipses

In Problems 1 – 4, graph the ellipse. Find (a) the vertices and (b) the foci of each ellipse.

1. $\dfrac{x^2}{25} + \dfrac{y^2}{4} = 1$

2. $\dfrac{x^2}{16} + \dfrac{y^2}{36} = 1$

1a. _____

1b. _____

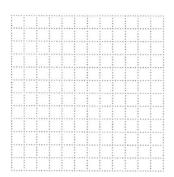
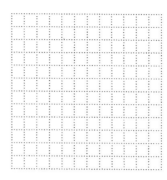

2a. _____

2b. _____

3. $\dfrac{x^2}{64} + y^2 = 1$

4. $9x^2 + y^2 = 81$

3a. _____

3b. _____

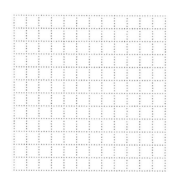
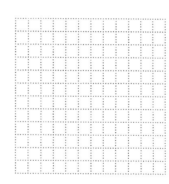

4a. _____

4b. _____

In Problems 5 – 8, find an equation for each ellipse.

5. center at (0, 0); focus at (2, 0); vertex at (5, 0)

5. _____

6. center at (0, 0); focus at (0, −1); vertex at (0, 5)

6. _____

7. foci at (0, ±2); vertices at (0, ±7)

7. _____

8. foci at (±6, 0); length of the major axis is 20

8. _____

In Problems 9 and 10, graph each ellipse.

9. $\dfrac{(x+8)^2}{81}+(y-3)^2=1$

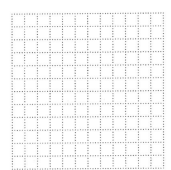

10. $9(x-3)^2+(y-4)^2=81$

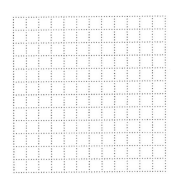

11. Consider the graph of the ellipse: $25x^2+150x+9y^2-72y+144=0$.

    (a) Write the equation of the ellipse in standard form.   11a. _____

    (b) Find the center.   11b. _____

    (c) Find the vertices.   11c. _____

    (d) Find the foci.   11d. _____

12. **London Bridge** An arch in the shape of the upper half of an ellipse is used to support London Bridge. The main span is 45.6 meters wide. Suppose that the center of the arch is 15 meters above the center of the river.

    (a) Write the equation for the ellipse in which the *x*-axis coincides with the water and the *y*-axis passes though the center of the arch.   12a. _____

    (b) Can a rectangular barge that is 20 meters wide and sits 12 meters above the surface of the water fit through the opening of the bridge?   12b. _____

Five-Minute Warm-Up 9.5
Hyperbolas

In Problems 1 and 2, solve the equation.

1. $y^2 = 16$

2. $(y + 3)^2 = 4$

1. _____

2. _____

In Problems 3 – 4, complete the square. Determine the number that must be added to the expression to make it a perfect square trinomial. Then factor the expression.

3. $x^2 + 8x$

4. $y^2 - \dfrac{4}{3}y$

3. _____

4. _____

5. Graph $y = \pm\dfrac{2}{3}x$. That is, graph $y = \dfrac{2}{3}x$ and $y = -\dfrac{2}{3}x$ on the same coordinate plane.

5. _____

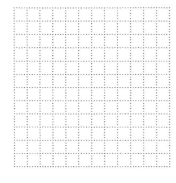

Guided Practice 9.5
Hyperbolas

Objective 1: Graph a Hyperbola Whose Center Is the Origin

1. A *hyperbola* is the collection all points in the plane the difference of whose distances from two fixed points is a positive constant.

(a) The two fixed points are called _____.

(b) The line containing these points, as well as the center and the vertices, is called the _____.

(c) The midpoints of the line segment joining the foci is the _____.

(d) The line through the center, perpendicular to the transverse axis, is called the _____.

(e) The branches of the hyperbola have turning points called _____.

2. The standard form of a hyperbola is either $\dfrac{x^2}{a^2} - \dfrac{y^2}{b^2} = 1$ or $\dfrac{y^2}{a^2} - \dfrac{x^2}{b^2} = 1$.
(See textbook Examples 1 and 2)

(a) In either of these equations, the first denominator is _____, whether it is larger or smaller.

(b) The distance from the center to the vertex is _____ units of length.

(c) Therefore, the length of the transverse axis is _____.

(d) The length of the conjugate axis is _____.

(e) If the first, or positive, term is $\dfrac{x^2}{a^2}$, the hyperbola opens _____.

(f) If the first, or positive, term is $\dfrac{y^2}{a^2}$, the hyperbola opens _____.

(g) The foci are c units from the center, on the transverse axis, where $c^2 =$ _____.

(h) For these two equations, the center of the hyperbola is _____.

3. Graph the hyperbola $\dfrac{x^2}{16} - \dfrac{y^2}{9} = 1$. *(See textbook Example 1)*

(a) Find the center.

3a. _____

(b) The equation is of the form $\dfrac{x^2}{a^2} - \dfrac{y^2}{b^2} = 1$, where $c^2 = a^2 + b^2$. Find the values of a and b.

3b. _____

(c) Find the value of c.  3c. _____

(d) Find the vertices and foci. 3d. _____

(e) Let $x = \pm c$ to locate points above and below the foci. 3e. _____

(f) Plot the vertices, foci, and the four points found in **(e)** to graph the hyperbola.

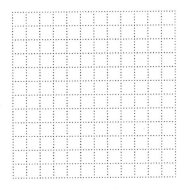

Objective 3: *Find the Asymptotes of a Hyperbola Whose Center Is the Origin*

4. In your own words, what is an *asymptote*?

5. The equations of the two asymptotes of the hyperbola can be found using the values of a and b as determined from the standard form. The equations for the asymptotes changes when the direction the hyperbola opening changes. Determine the equations of the asymptotes for each hyperbola:

Hyperbola	Equation of Asymptotes
$\dfrac{x^2}{a^2} - \dfrac{y^2}{b^2} = 1$	**(a)** $y = \pm$
$\dfrac{y^2}{a^2} - \dfrac{x^2}{b^2} = 1$	**(b)** $y = \pm$

Do the Math Exercises 9.5
Hyperbolas

In Problems 1 – 4, graph each hyperbola. Find (a) the vertices and (b) the foci.

1. $\dfrac{x^2}{9} - \dfrac{y^2}{16} = 1$

2. $\dfrac{y^2}{81} - \dfrac{x^2}{9} = 1$

1a. _____

1b. _____

2a. _____

2b. _____

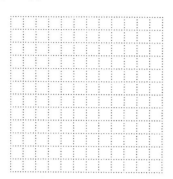
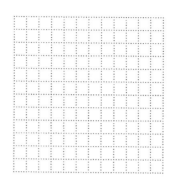

3. $x^2 - 9y^2 = 36$

4. $4y^2 - 9x^2 = 36$

3a. _____

3b. _____

4a. _____

4b. _____

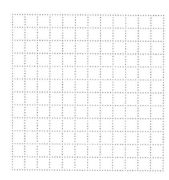
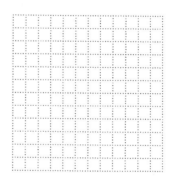

In Problems 5 – 8, find the equation for the hyperbola described.

5. center at (0, 0); focus at (–4, 0); and vertex (–1, 0)

5. _____

6. vertices at (0, 6) and (0, –6); focus at (0, 8)

6. _____

7. vertices at (0, –4) and (0, 4); asymptote the line $y = 2x$

7. _____

8. foci at (–9, 0) and (9, 0); asymptote the line $y = –3x$

8. _____

In Problems 9 – 14, identify the graph of each equation as a circle, parabola, ellipse, or hyperbola.

9. $x^2 + 4y = 4$

10. $3y^2 - x^2 = 9$

9. _____

10. _____

11. $y^2 - 4y - 4x^2 + 8x = 4$

12. $4x^2 + 8x + 4y^2 - 4y - 12 = 0$

11. _____

12. _____

13. $4y^2 + 3x - 16y + 19 = 0$

14. $9x^2 + 4y^2 - 18x + 8y - 23 = 0$

13. _____

14. _____

15. Consider the hyperbola $\dfrac{y^2}{25} - \dfrac{x^2}{4} = 1$.

    (a) Find the vertices.

15a. _____

    (b) Find the foci.

15b. _____

16. Consider the hyperbola $\dfrac{x^2}{16} - \dfrac{y^2}{36} = 1$. Find the equation of the asymptotes.

16. _____

Five-Minute Warm-Up 9.6
Systems of Nonlinear Equations

In Problems 1 and 2, solve the system using substitution.

1. $\begin{cases} y = \dfrac{3}{4}x \\ x - 4y = -4 \end{cases}$

2. $\begin{cases} -x + 3y = 4 \\ 2x - 6y = -8 \end{cases}$

1. _____

2. _____

In Problems 3 and 4, solve the system using elimination.

3. $\begin{cases} 6x - 5y = 1 \\ 8x - 2y = -22 \end{cases}$

4. $\begin{cases} 6x - 4y = -5 \\ -12x + 8y = 2 \end{cases}$

3. _____

4. _____

Guided Practice 9.6
Systems of Nonlinear Equations

Objective 1: Solve a System of Nonlinear Equations Using Substitution

1. Solve the following system of equations using substitution: $\begin{cases} y = x - 4 & (1) \\ x^2 + y^2 = 16 & (2) \end{cases}$

(See textbook Example 1)

Step 1: Graph each equation in the system.

Graph of equation (1) is what? **(a)** _____

Graph of equation (2) is what? **(b)** _____

Graph each equation in the system: **(c)**

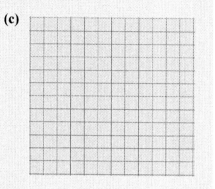

Based on (c), how many solutions will this system have? **(d)** _____

Step 2: Solve equation (1) for y.

This is already done.

Step 3: Substitute the expression for y into equation (2).

Equation (2): **(e)** _____

Substitute the expression for y into equation (2): **(f)** _____

Simplify: **(g)** _____

Step 4: Solve for x.

Factor: **(h)** _____

Zero-Product Property: **(i)** $x =$ _____ or $x =$ _____

Step 5: Use your values from (i) and equation (1) to determine the ordered pairs that satisfy the system.

Equation (1): **(j)** _____

Substitute your first value of x: **(k)** _____

Solve for y: **(l)** _____

Equation (1): **(m)** _____

Substitute your second value of x: **(n)** _____

Solve for y: **(o)** _____

Step 6: Check

State the solution set: **(p)** _____

Objective 2: Solve a System of Nonlinear Equations Using Elimination

2. Solve the following system of equations by elimination: $\begin{cases} x^2 + y^2 = 4 & (1) \\ x^2 + 4y^2 = 16 & (2) \end{cases}$ *(See textbook Example 3)*

Step 1: Graph each equation in the system.

Graph of equation (1) is what? **(a)** _____

Graph of equation (2) is what? **(b)** _____

Graph each equation in the system: **(c)**

Based on (c), how many solutions will this system have? **(d)** _____

Step 2: We want a pair of coefficients to be additive inverses so that when we add equation (1) and equation (2), one of the variables will be eliminated.

Let's eliminate x^2. What will equation (1) need to be multiplied by in order to eliminate x^2? **(e)** _____

Multiply equation (1) by the value determined in (e) and write the system: **(f)** $\begin{cases} \underline{\hspace{3cm}} & (1) \\ \underline{\hspace{3cm}} & (2) \end{cases}$

Step 3: Add equations (1) and (2) to eliminate x^2. Solve the resulting equation for y.

Add: **(g)** _____

Divide by 3: **(h)** _____

Use the Square Root Property: **(i)** $y = $ _____

Step 4: Solve for x using either equation (1) or equation (2). We will use your values from (i) and equation (1) to determine the ordered pairs that satisfy the system.

Equation (1): **(j)** _____

Substitute your first value of y: **(k)** _____

Solve for x: **(l)** _____

Equation (1): **(m)** _____

Substitute your second value of y: **(n)** _____

Solve for x: **(o)** _____

Step 5: Check

State the solution set: **(p)** _____

Sullivan/Struve, *Intermediate Algebra*, 3e
Copyright © 2014 Pearson Education, Inc.

Name:
Instructor:

Date:
Section:

Do the Math Exercises 9.6
Systems of Nonlinear Equations

In Problems 1 – 4, solve the system of nonlinear equations by substitution.

1. $\begin{cases} y = x^3 + 2 \\ y = x + 2 \end{cases}$

2. $\begin{cases} y = \sqrt{100 - x^2} \\ x + y = 14 \end{cases}$

1. _____

2. _____

3. $\begin{cases} x^2 + y^2 = 16 \\ y = x^2 - 4 \end{cases}$

4. $\begin{cases} xy = 1 \\ x^2 - y = 0 \end{cases}$

3. _____

4. _____

In Problems 5 – 8, solve the system of nonlinear equations by elimination.

5. $\begin{cases} x^2 + y^2 = 8 \\ x^2 + y^2 + 4y = 0 \end{cases}$

6. $\begin{cases} 4x^2 + 16y^2 = 16 \\ 2x^2 - 2y^2 = 8 \end{cases}$

5. _____

6. _____

7. $\begin{cases} 2x^2 + y^2 = 18 \\ x^2 - y^2 = 9 \end{cases}$

8. $\begin{cases} 2x^2 - 5x + y = 12 \\ 14x - 2y = -16 \end{cases}$

7. _____

8. _____

In Problems 9 – 12, solve the system of nonlinear equations by any method.

9. $\begin{cases} y = x^2 + 4x + 5 \\ x - y = 9 \end{cases}$

10. $\begin{cases} x^2 + y^2 = 25 \\ x^2 - y^2 = 25 \end{cases}$

9. _____

10. _____

11. $\begin{cases} 9x^2 + 4y^2 = 36 \\ x^2 + (y-7)^2 = 4 \end{cases}$

12. $\begin{cases} x^2 + y^2 = 65 \\ y = -x^2 + 9 \end{cases}$

11. _____

12. _____

13. **Fun with Numbers** The sum of two numbers is 8. The sum of their squares is 160. Find the numbers.

13. _____

14. **Perimeter and Area of a Rectangle** The perimeter of a rectangle is 64 meters. The area of the rectangle is 240 square feet. Find the dimensions of the rectangle.

14. _____

Five-Minute Warm-Up 10.1
Sequences

1. If $f(x) = -2x^2 - 3x$, find the function value.

　(a) $f(1)$　　　　　　　　　　　**(b)** $f(-3)$

1a. _____

1b. _____

2. Evaluate the expression $(-1)^n (2n - 3)$ for each of the following.

　(a) $n = 1$　　　**(b)** $n = 2$　　　**(c)** $n = 3$　　　**(d)** $n = 4$

2a. _____

2b. _____

2c. _____

2d. _____

3. Evaluate the expression $\left(-\dfrac{1}{3}\right)^{n+1}$ for each of the following.

　(a) $n = 1$　　　　　**(b)** $n = 2$　　　　　**(c)** $n = 3$

3a. _____

3b. _____

3c. _____

4. If $f(x) = \dfrac{1}{2x}$, find $f(1) + f(2) + f(3) + f(4)$.

4. _____

Guided Practice 10.1
Sequences

1. A *sequence* is a function whose domain is the set of positive integers.

    (a) The numbers in the ordered list are called _____ of the sequence and we separate each entry on the list from the next entry by a comma.

    (b) Sequences can be either infinite or finite. If the list does not end, it is called *infinite* and we use three dots, called _____ , to indicate that the pattern continues indefinitely.

    (c) By contrast, a _____ sequence has a countable number of terms.

Objective 1: Write the First Few Terms of a Sequence

2. We use the notation a_6 to mean the sixth term of the sequence. The formula for the *n*th term, or *general term*, of the sequence is denoted

                                                                   2. _____

3. Write the first five terms of the sequence $\{a_n\} = \left\{\dfrac{2^n - 1}{n}\right\}$. *(See textbook Example 1)*

                                                          3. _____

Objective 2: *Find a Formula for the nth Term of a Sequence*

4. Sometimes a sequence is indicated by an observed pattern and it is our job to find the pattern. Try subtracting successive terms to find a constant difference or dividing successive terms to find a constant ratio. Sometimes the pattern is neither of these, but rather something you can recognize such as perfect squares.

Find the formula for the *n*th term of the sequence: *(See textbook Example 3)*

(a) 3, 6, 9, 12, …

(b) $\dfrac{1}{6}, -\dfrac{1}{36}, \dfrac{1}{216}, -\dfrac{1}{1296}, \dots$

4a. _____

4b. _____

Objective 3: *Use Summation Notation*

5. We use *summation notation* to indicate that the terms of the sequence should be added. The *index* of summation can be any variable, but typically we use *i*. This tells you where to start the sum and where to end. When there are a finite number of terms to be added, the sum is called a *partial sum*.

Consider the partial sum: $\displaystyle\sum_{i=1}^{4}\left(i^2 + 3\right)$. We substitute the values $i =$ ___, $i =$ ___, $i =$ ___, $i =$ ___ into the

formula $i^2 + 3$ to get the terms _____, _____, _____, _____. The sum is _____. *(See textbook Example 4)*

6. Write out the sum and determine its value: $\displaystyle\sum_{k=0}^{5}\left(2k - 5\right)$ _____

7. Express the sum using summation notation: $0 + 2 + 4 + 6 + 8 + 10 + 12$ *(See textbook Example 5)*

7. _____

Name: Date:
Instructor: Section:

Do the Math Exercises 10.1
Sequences

In Problems 1 – 4, write down the first five terms of each sequence.

1. $\{n-4\}$

2. $\left\{\dfrac{n+4}{n}\right\}$

1._____

2._____

3. $\{3^n-1\}$

4. $\left\{\dfrac{n^2}{2}\right\}$

3._____

4._____

In Problems 5 – 8, find the nth term of each sequence suggested by the pattern.

5. $5, 10, 15, 20, \ldots$

6. $\dfrac{1}{2}, 1, \dfrac{3}{2}, 2, \dfrac{5}{2}, \ldots$

5._____

6._____

7. $0, 7, 26, 63, \ldots$

8. $1, -\dfrac{1}{2}, \dfrac{1}{4}, -\dfrac{1}{8}, \ldots$

7._____

8._____

In Problems 9 - 14, determine the value of the sum.

9. $\displaystyle\sum_{i=1}^{5}(3i+2)$

10. $\displaystyle\sum_{i=1}^{4}\dfrac{i^3}{2}$

9._____

10._____

11. $\displaystyle\sum_{k=1}^{4}3^k$

12. $\displaystyle\sum_{k=1}^{8}\left[(-1)^k \cdot k\right]$

11._____

12._____

13. $\displaystyle\sum_{j=1}^{8}2$

14. $\displaystyle\sum_{j=5}^{10}(k+4)$

13._____

14._____

In Problems 15 – 18, express each sum using summation notation.

15. $1 + 3 + 5 + \dots + 17$

16. $1 + \dfrac{1}{2} + \dfrac{1}{4} + \dots + \dfrac{1}{2^{15}}$

15. _____

16. _____

17. $\dfrac{2}{3} - \dfrac{4}{9} + \dfrac{8}{27} + \dots + (-1)^{15+1}\left(\dfrac{2}{3}\right)^{15}$

18. $3 + 3 \cdot \dfrac{1}{2} + 3 \cdot \dfrac{1}{4} + \dots + 3 \cdot \left(\dfrac{1}{2}\right)^{11}$

17. _____

18. _____

19. The Future Value of Money Suppose that you place \$5,000 into a company 401(k) plan that pays 8% interest compounded monthly. The balance in the account after n months is given by $a_n = 5,000\left(1 + \dfrac{0.08}{12}\right)^n$.

(a) Find the value in the account after 1 month.

19a. _____

(b) Find the value in the account after 1 year.

19b. _____

(c) Find the value in the account after 10 years.

19c. _____

Sullivan/Struve, *Intermediate Algebra*, 3e
Copyright © 2014 Pearson Education, Inc.

Five-Minute Warm-Up 10.2
Arithmetic Sequences

1. Determine the slope of $y = 4x + 5$.

1. _____

2. If $g(x) = \dfrac{2}{3}x - 1$, find $g\left(-\dfrac{6}{5}\right)$.

2. _____

3. Solve the following systems of linear equations:

(a) $\begin{cases} 3x - 5y = 22 \\ x - y = 10 \end{cases}$

(b) $\begin{cases} x - 2y = 9 \\ 2x + y = -2 \end{cases}$

3a. _____

3b. _____

4. Evaluate the expression: $\dfrac{9}{2}\left(-\dfrac{4}{3} + \left(-\dfrac{20}{3}\right)\right)$

4. _____

Guided Practice 10.2
Arithmetic Sequences

● **Objective 1: Determine Whether a Sequence Is Arithmetic**

1. If there is a constant difference between the successive terms, the sequence is called an *arithmetic sequence*.

    (a) In formulas for arithmetic sequences, we label the common difference _____.

    (b) We label the first term _____.

2. Determine if the sequence 5, 8, 11, 14, … is arithmetic. If it is, determine the first term and the common difference. *(See textbook Example 1)*

                    2._____ _____

3. Show that the sequence $\{a_n\} = \{2 + n^2\}$ is not arithmetic by listing the first six terms and calculating the difference between successive terms. *(See textbook Example 3)*

                    3. _____

●

Objective 2: Find a Formula for the nth Term of an Arithmetic Sequence

4. The *n*th term of an arithmetic sequence whose first term is a_1 and whose common difference is d, is determined by the formula: $a_n = a_1 + (n-1)d$.

(a) Find a formula for the *n*th term of the arithmetic sequence whose first term is -3 and whose common difference is 5. *(See textbook Example 4)*   4a. _____

(b) Find the 8th term of this sequence.   4b. _____

5. The 5th term of an arithmetic sequence is 24, and the 11th term is 42. *(See textbook Example 5)*

(a) Write and solve a system of linear equations to find the first term and the common difference.   5a. _____

● **(b)** Give a formula for the *n*th term.   5b. _____

Objective 3: Find the Sum of an Arithmetic Sequence

6. Let $\{a_n\}$ be an arithmetic sequence with first term a_1 and common difference d. The sum

S_n of the first n terms of $\{a_n\}$ is $S_n = \dfrac{n}{2}(a_1 + a_n)$.

Find the sum of the first 10 terms of the arithmetic sequence 12, 16, 20, 24, ... *(See textbook Example 6)*

(a) Write a formula for the nth term.   6a. _____

(b) Use the formula to find the 10th term of the sequence.   6b. _____

(c) Find the sum of the first 10 terms of the arithmetic sequence.   6c. _____

7. Find the sum of the first 10 terms of the arithmetic sequence $\{-10n + 13\}$. *(See textbook Example 7)*

(a) Use the formula to find the 1st and 10th term of the sequence.   7a. _____

(b) Find the sum of the first 10 terms of the arithmetic sequence.   7b. _____

Do the Math Exercises 10.2
Arithmetic Sequences

In Problems 1 – 2, find the common difference and write out the first four terms.

1. $\{10n + 1\}$

2. $\left\{\dfrac{1}{4}n + \dfrac{3}{4}\right\}$

1. _____

2. _____

In Problems 3 – 5, find a formula for the nth term of the arithmetic sequence whose first term is $\underline{a}$ and common difference $\underline{d}$ is given. What is the fifth term?

3. $a = 8, d = 3$

4. $a = 12, d = -3$

3. _____

4. _____

5. $a = -3; d = \dfrac{1}{2}$

5. _____

In Problems 6 – 8, write a formula for the nth term of each arithmetic sequence. Use the formula to find the 20th term of the sequence.

6. $-5, -1, 3, 7, \ldots$

7. $20, 14, 8, 2, \ldots$

6. _____

7. _____

8. $10, \dfrac{19}{2}, 9, \dfrac{17}{2}, \ldots$

8. _____

In Problems 9 – 12, find the formula for the nth term of an arithmetic sequence using the given information.

9. 5th term is 7; 9th term is 19

10. 2nd term is –9; 8th term is 15

9. _____

10. _____

11. 6th term is –8; 12th term is –38

12. 5th term is 5; 13th term is 7

11. _____

12. _____

13. Find the sum of the first 40 terms of the sequence 1, 8, 15, 22, ...

13. _____

14. Find the sum of the first 75 terms of the sequence –9, –5, –1, 3 ...

14. _____

15. Find the sum of the first 50 terms of the sequence 12, 4, –4, –12, ...

15. _____

16. Find the sum of the first 80 terms of the arithmetic sequence $\{2n - 13\}$.

16. _____

17. Find the sum of the first 35 terms of the arithmetic sequence $\{-6n + 25\}$.

17. _____

18. Find the sum of the first 28 terms of the arithmetic sequence $\left\{ 7 - \dfrac{3}{2}n \right\}$.

18. _____

19. Find x so that $2x$, $3x + 2$, and $5x + 3$ are consecutive terms of an arithmetic sequence.

19. _____

20. The Theater An auditorium has 40 seats in the first row and 25 rows in all. Each successive row contains 2 additional seats. How many seats are in the auditorium?

20. _____

Sullivan/Struve, *Intermediate Algebra,* 3e
Copyright © 2014 Pearson Education, Inc.

Five-Minute Warm-Up 10.3
Geometric Sequences and Series

1. If $f(x) = \left(\dfrac{2}{3}\right)^x$, find each of the following.

    (a) $f(1)$ **(b)** $f(2)$ **(c)** $f(3)$ 1a. _____

1b. _____

1c. _____

2. If $g(n) = 3n^2$, find each of the following.

    (a) $g(1)$ **(b)** $g(2)$ **(c)** $g(3)$ 2a. _____

2b. _____

2c. _____

3. Simplify each expression.

    (a) $\dfrac{24x^5}{15x^2}$ **(b)** $\left(-3r^4\right)^2$ 3a. _____

3b. _____

4. Evaluate the expression: $\dfrac{\frac{3}{2}}{1 - \frac{1}{4}}$ 4. _____

Guided Practice 10.3
Geometric Sequences and Series

Objective 1: Determine Whether a Sequence Is Geometric

1. If there is a constant ratio between the successive terms, the sequence is called a *geometric sequence*.

    (a) In formulas for geometric sequences, we label the common ratio _____.

    (b) We label the first term _____.

2. Determine if the sequence $36, 18, 9, \dfrac{9}{2}, \ldots$ is geometric. If it is, determine the first term and the common ratio. *See Example 1*

2. _____

3. Show that the sequence $\{b_n\} = \left\{\left(\dfrac{2}{5}\right)^{n-1}\right\}$ is geometric by listing the first four terms and calculating the ratio between successive terms. *(See textbook Example 3)*

3. _____

Objective 2: Find a Formula for the nth Term of a Geometric Sequence

4. The *n*th term of a geometric sequence whose first term is a and whose common ratio is r, is determined by the formula: $a_n = a_1 r^{n-1}; r \neq 0$.

(a) Find a formula for the *n*th term of the geometric sequence: $2, -\dfrac{2}{3}, \dfrac{2}{9}, -\dfrac{2}{27}, \ldots$.   4a. _____
(See textbook Example 4)

(b) Find the 8th term of this sequence.   4b. _____

Objective 3: Find the Sum of a Geometric Sequence

5. Let $\{a_n\}$ be a geometric sequence with first term a_1 and common ratio r, where $r \neq 0, r \neq 1$.

The sum S_n of the first n terms of $\{a_n\}$ is $S_n = a_1 \cdot \dfrac{1 - r^n}{1 - r}$; $r \neq 0, r \neq 1$.

(a) Find the first term and the common ratio, r, of the geometric sequence 6, 24, 96, 384, ...
(See textbook Example 5)

5a. _____

(b) Find the sum of the first 10 terms of this sequence.

5b. _____

Objective 4: Find the Sum of a Geometric Series

6. An infinite sum of the terms of a geometric sequence is called a *geometric series*. We can find the sum of the series with the formula: $\displaystyle\sum_{n=1}^{\infty} a_1 r^{n-1} = \dfrac{a_1}{1 - r}$ provided that $-1 < r < 1$.

(a) Find the first term and the common ratio, r, of the geometric series: $6 - 2 + \dfrac{2}{3} - \dfrac{2}{9} + \ldots$

(See textbook Example 7)

6a. _____

(b) Find the sum of this series.

6b. _____

Objective 5: Solve Annuity Problems

7. If P represents the deposit in dollars made at each payment period for an annuity at i percent interest per payment period, the amount A of the annuity after n payment periods is: $A = P \cdot \dfrac{(1 + i)^n - 1}{i}$.

Retirement Raymond is planning on retiring in 15 years, so he contributes $1,500 into his IRA every 6 months (semiannually). What will be the value of the IRA when Raymond retires if earns 10% interest compounded semiannually? *(See textbook Example 10)*

7. _____

Do the Math Exercises 10.3
Geometric Sequences and Series

In Problems 1 – 3, find the common ratio and write out the first four terms of each geometric sequence.

1. $\{(-2)^n\}$

2. $\left\{\dfrac{2^n}{3}\right\}$

1. _____

2. _____

3. $\left\{\dfrac{3^{-n}}{2^{n-1}}\right\}$

3. _____

In Problems 4 – 7, determine whether the given sequence is arithmetic, geometric, or neither.

4. $\{8 - 3n\}$

5. $\{n^2 - 2\}$

4. _____

5. _____

6. $100, 20, 4, \dfrac{4}{5}, \ldots$

7. $5, -2, 3, -1, 2, \ldots$

6. _____

7. _____

In Problems 8 and 9, find a formula for the nth term of the geometric sequence whose first term and common ratio are given. Use the formula to find the 8ᵗʰ term.

8. $a_1 = 30, r = \dfrac{1}{3}$

9. $a_1 = 1, r = -4$

8. _____

9. _____

In Problems 10 – 12, find the indicated term of each geometric sequence.

10. 12ᵗʰ term of 1, 3, 9, 27, …

11. 8ᵗʰ term of 10, –20, 40, –80, …

10. _____

11. _____

12. 10ᵗʰ term of 0.4, 0.04, 0.004, 0.0004, …

12. _____

In Problems 13 and 14, find the sum of each geometric series.

13. $3 + 9 + 27 + \ldots + 3^{10}$

14. $\sum\limits_{n=1}^{12}\left[5 \cdot 2^n\right]$

13. _____

14. _____

In Problems 15 – 17, find the sum of each infinite geometric series.

15. $1 + \dfrac{1}{3} + \dfrac{1}{9} + \ldots$

16. $12 - 3 + \dfrac{3}{4} - \dfrac{3}{16} + \ldots$

15. _____

16. _____

17. $\sum\limits_{n=1}^{\infty}\left(10 \cdot \left(\dfrac{1}{3}\right)^n\right)$

17. _____

In Problems 18 and 19, express each repeating decimal as a fraction in lowest terms.

18. $0.\overline{3}$

19. $0.\overline{45}$

18. _____

19. _____

20. Depreciation of a Car Suppose that you have just purchased a Chevy Impala for $16,000. Historically, the car depreciates by 10% each year, so that next year the car is worth $16,000(0.9). What will the value of the car be after you have owned it for four years?

20. _____

Sullivan/Struve, *Intermediate Algebra*, 3e
Copyright © 2014 Pearson Education, Inc.

Five-Minute Warm-Up 10.4
The Binomial Theorem

In Problems 1 – 4, find the product.

1. $(x + 2)^2$

2. $(y - 3)^2$

1. _____

2. _____

3. $(4x - 5y)^2$

4. $\left(3n + \dfrac{2}{3}\right)^2$

3. _____

4. _____

5. Simplify: $\dfrac{7 \bullet 6 \bullet 5 \bullet 4 \bullet 3 \bullet 2}{3 \bullet 2 \bullet 4 \bullet 3 \bullet 2}$

5. _____

Guided Practice 10.4
The Binomial Theorem

Objective 1: Compute Factorials

1. If $n \geq 0$ is an integer, the **factorial symbol** $n!$ (read "n factorial") is defined as:

$$0! = 1; \ 1! = 1; \ n! = n(n-1)(n-2) \cdot \ldots \cdot 3 \cdot 2 \cdot 1 \text{ for } n \geq 2.$$

Evaluate each of the following. *(See textbook Example 1)*

(a) $\dfrac{10!}{4!}$

(b) $\dfrac{6!}{2!(6-2)!}$

1a. _____

1b. _____

Objective 2: Evaluate a Binomial Coefficient

2. If j and n are integers with $0 \leq j \leq n$, the symbol $\dbinom{n}{j}$ (read "n choose j") is defined as

$$\binom{n}{j} = \frac{n!}{j!(n-j)!}.$$

Evaluate each of the following. *(See textbook Example 2)*

(a) $\dbinom{5}{3}$

(b) $\dbinom{13}{7}$

2a. _____

2b. _____

Objective 3: Expand a Binomial

3. Multiplying binomials can become unwieldy when the exponents become large. We now introduce techniques for binomial expansion. We can find the binomial coefficients in a binomial expansion using Pascal's Triangle or the Binominal Theorem, which states that for any positive integer n,

$$(x+a)^n = \binom{n}{0}x^n + \binom{n}{1}ax^{n-1} + \binom{n}{2}a^2x^{n-2} + \ldots + \binom{n}{j}a^j x^{n-j} + \ldots + \binom{n}{n}a^n.$$

Expand $(p+2)^4$ using the Binomial Theorem. *(See textbook Example 3)*

Sullivan/Struve, *Intermediate Algebra,* 3e

Do the Math Exercises 10.4
The Binomial Theorem

In Problems 1 – 4, evaluate each expression.

1. $5!$

2. $\dfrac{6!}{2!}$

1. _____

2. _____

3. $\dfrac{10!}{2! \cdot 8!}$

4. $0!$

3. _____

4. _____

In Problems 5 – 8, evaluate each expression.

5. $\dbinom{5}{3}$

6. $\dbinom{7}{5}$

5. _____

6. _____

7. $\dbinom{50}{49}$

8. $\dbinom{1000}{1000}$

7. _____

8. _____

In Problems 9 – 16, expand each expression using the Binomial Theorem.

9. $(x-1)^4$

10. $(x+5)^5$

9. _____

10. _____

11. $(2q + 3)^4$

12. $(3w - 4)^4$

11. _____

12. _____

13. $(y^2 - 3)^4$

14. $(3b^2 + 2)^5$

13. _____

14. _____

15. $(p - 3)^6$

16. $(3x^2 + y^3)^4$

15. _____

16. _____

17. Use the Binomial Theorem to find the numerical value of $(1.001)^4$ correct to five decimal places. [Hint: $(1.001)^5 = (1 + 10^{-3})^5$].

17. _____

Chapter R Answers

Section R.2

Five-Minute Warm-Up   1. $\subseteq$ **2.** $\subset$ **3.** $\varnothing$ **4.** $\in$ **5.** $\notin$ **6.** $>$ **7.** $<$ **8.** $\geq$ **9.** $\leq$ **10.** $x > 0$

11. $x < 0$ **12 – 17.** Answers may vary   **12.** $\{1, 2, 3...\}$ **13.** $\{0, 1, 2...\}$ **14.** $\{... -2, -1, 0, 1, 2...\}$

15. $\left\{-\dfrac{1}{3}, 0.24, 0.\overline{7}\right\}$ **16.** $\left\{\sqrt{2}, \sqrt{13}, \pi\right\}$ **17.** $\left\{-4, 0, 2, \pi, \sqrt{6}, \dfrac{9}{4}\right\}$ **18.** $0.111...$; nonterminating

19. 0.375; terminating

Guided Practice   1. In order for A to be a proper subset of B, every element of A is an element of B, but there are elements of B that are not in A.   **2.** yes   **3.** yes   **4.** There are elements of B which are not elements of A.   **5.** $\{a\}, \{b\}, \{c\}, \{a, b\}, \{a, c\}, \{b, c\}, \{a, b, c\}, \varnothing$ **6a.** TRUE   **6b.** TRUE   **6c.** FALSE   **6d.** TRUE   **6e.** TRUE   **6f.** FALSE   **6g.** TRUE   **6h.** FALSE   **6i.** FALSE   **6j.** TRUE   **6k.** TRUE   **6l.** FALSE

7a. $\left\{\text{undefined}, -1, 1.333..., \pi, -2, 100,000, -\dfrac{5}{4}, 3, 0, \sqrt{5}, 0.25, \text{not a real number}\right\}$ **7b.** $\left\{100,000, \sqrt{9}\right\}$

7c. $\left\{100,000, \sqrt{9}, \dfrac{0}{2}\right\}$ **7d.** $\left\{100,000, \sqrt{9}, \dfrac{0}{2}, -1, \dfrac{-6}{3}\right\}$ **7e.** $\left\{100,000, \sqrt{9}, \dfrac{0}{2}, -1, \dfrac{-6}{3}, 1.\overline{3}, -\dfrac{5}{4}, 0.25\right\}$

7f. $\left\{\pi, \sqrt{5}\right\}$ **7g.** $\left\{100,000, \sqrt{9}, \dfrac{0}{2}, -1, \dfrac{-6}{3}, 1.\overline{3}, -\dfrac{5}{4}, 0.25, \pi, \sqrt{5}\right\}$

Do the Math   1. $\{-3, -2, -1, 0, 1, 2, 3, 4, 5\}$ **2.** $\varnothing$ or $\{\ \}$ **3.** TRUE   **4.** TRUE   **5.** TRUE

6. TRUE   **7.** $\in$ **8.** $\notin$ **9a.** 4   **9b.** $-5, 4$ **9c.** $-5, 4, \dfrac{4}{3}, -\dfrac{7}{5}, 5.\overline{1}$ **9d.** π **9e.** $-5, 4, \dfrac{4}{3}, -\dfrac{7}{5}, 5.\overline{1}, \pi$ **10a.** 13

10b. $0, 13$ **10c.** $-4.5656..., 0, 2.43, 13, \dfrac{8}{7}$ **10d.** $\sqrt{2}$ **10e.** $-4.5656..., 0, 2.43, 13, \dfrac{8}{7}, \sqrt{2}$ **11a.** -9.99

11b. -10.00 **12.** See Graphing Answer Section   **13.** $>$ **14.** $>$ **15.** $-\dfrac{37}{100}$ **16.** 1.143 **17.** 1.142

18. $0.666...$; repeating   **19.** -0.16; terminating   **20a.** no   **20b.** no   **20c.** The set of real numbers is the union of the set of rational numbers and the set of irrational numbers. Therefore, all real numbers are either rational or irrational. Since decimals cannot be simultaneously repeating and nonrepeating, there are no real numbers that are both rational and irrational.

Section R.3

Five-Minute Warm-Up   1. The distance a number is from zero. The absolute value of a number can never be negative.   **2.** Answer when two numbers are multiplied.   **3.** Answer when two numbers are divided.   **4.** Answer when two numbers are added.   **5.** Answer when two numbers are subtracted.   **6.** 0   **7.** 1

8. undefined   **9a.** 4   **9b.** 12   **9c.** 3   **10a.** 20   **10b.** 5   **10c.** 0.25   **11.** $\dfrac{5}{4}$ **12.** 1   **13.** $\dfrac{3}{10}$ **14.** $\dfrac{15}{8}$

Guided Practice   1a. positive   **1b.** negative   **2.** the same as the sign of the number with the larger absolute value   **3.** See Graphing Answer Section   **4a.** $12 = 2 \cdot 2 \cdot 3$ **4b.** $18 = 2 \cdot 3 \cdot 3$ **4c.** 2, 3   **4d.** 2, 3

4e. $2 \cdot 2 \cdot 3 \cdot 3 = 36$ **4f.** 21   **4g.** $\dfrac{2}{2}$; 10   **5a-k.** Answers will vary; **5a.** $a + 0 = a$ **5b.** $a \cdot 1 = a$

5c. $a + (-a) = 0$ **5d.** $a \cdot \dfrac{1}{a} = 1$ **5e.** $-(-a) = a$ **5f.** $a + b = b + a$ **5g.** $a \cdot b = b \cdot a$ **5h.** $a \cdot 0 = 0$

5i. $(a + b) + c = a + (b + c)$ **5j.** $(a \cdot b) \cdot c = a \cdot (b \cdot c)$ **5k.** $a \cdot (b + c) = a \cdot b + a \cdot c$ **6a.** $6x - 10$
6b. $-36y + 12$ **6c.** $20n + 8$ **6d.** $4x - 16$

Do the Math   1a. $\dfrac{1}{5}$ **1b.** -5 **2a.** -10 **2b.** $\dfrac{1}{10}$ **3.** $-2x - 10$ **4.** $\dfrac{5}{2}$ **5.** 16   **6.** -105 **7.** $\dfrac{4}{5}$

8. $-\dfrac{14}{3}$ **9.** 12   **10.** -2 **11.** $-\dfrac{481}{360}$ **12.** $\dfrac{1}{12}$ **13.** 5.1   **14.** 16   **15.** $-\dfrac{13}{2}$ **16.** $-\dfrac{7}{8}$ **17.** Associative of Addition   **18.** Multiplicative Inverse   **19.** Multiplicative Identity   **20.** Commutative of Multiplication   **21.** 543 feet   **22a.** $a \cdot 0 = 0$ **22b.** $a \cdot \big(b + (-b)\big) = 0$ **22c.** $ab + a(-b) = 0$

Do the Math **22d.** $ab < 0$ since the product of a negative and a positive is negative; $a(-b)$ must be positive so that $ab + a(-b) = 0$.

Section R.4

Five -Minute Warm-Up **1a.** base **1b.** exponent **1c.** $4 \cdot 4 \cdot 4 = 64$ **2a.** squared **2b.** $(3x)^2$ **3a.** cubed **3b.** $(y + 1)^3$ **4a.** 5 **4b.** $-2x$ **5.** $\dfrac{-27}{8}$ **6.** $-\dfrac{23}{18}$

Guided Practice **1a.** 36 **1b.** 36 **1c.** -36 **1d.** -36 **1e.** $-\dfrac{64}{125}$ **1f.** $\dfrac{256}{625}$ **2.** positive **3.** negative

4a. Parentheses **4b.** Exponents **4c.** Multiply or Divide **4d.** Add or Subtract **5a.** $\dfrac{3 \cdot 4}{-9 - 6 \cdot 3}$ **5b.** $\dfrac{12}{-9 - 18}$

5c. $-\dfrac{12}{-27}$ **5d.** $-\dfrac{4}{9}$

Do the Math **1.** 13 **2.** 31 **3.** -32 **4.** 33 **5.** 17 **6.** 3 **7.** 3 **8.** -4 **9.** $\dfrac{3}{2}$ **10.** 60 **11.** $\dfrac{1}{4}$

12. 180 **13.** -31 **14.** -46 **15.** 24 **16.** 8 **17.** $\dfrac{1}{2}$ **18.** $\dfrac{2}{3}$ **19.** ≈ 603.19 in.^{2} **20a.** $3 + 5 \cdot (6 - 3) = 18$

20b. $(3 + 5) \cdot (6 - 3) = 24$ **21.** We cannot use the Reduction Property across addition.

Section R.5

Five-Minute Warm-Up **1a.** $-10x + 15$ **1b.** $36a + 27$ **1c.** $4n - 10$ **1d.** $-2z + \dfrac{6}{5}$ **2a.** -8 **2b.** $\dfrac{2}{3}$

2c. 28 **2d.** $-\dfrac{11}{18}$

Guided Practice **1a.** $-2x$ **1b.** $\dfrac{y}{2} - 6$ or $\dfrac{1}{2}y - 6$ **1c.** $\dfrac{m}{8}$ **1d.** $25 - x$ **2.** 8 **3a.** 0 **3b.** $6x^2$ **4a.** $3x - 1$

4b. $2x + 16$ **5.** domain **6.** $x = 0$ **7.** the denominator to equal zero **8a.** $0, -2, 4$ **8b.** 3, 5, 0

Do the Math **1.** 2 **2.** -33 **3.** 11 **4.** undefined **5.** $7y$ **6.** $-x + 1$ **7.** $\dfrac{17}{30}y$ **8.** $-4x^2 + 3x$

9. $9y + 9$ **10.** 3 **11.** $12x - 4$ **12.** $-6w + 11$ **13.** $\dfrac{82}{15}y + \dfrac{44}{15}$ **14.** $0.35x - 17.38$ **15.** $10 - y$ **16.** $\dfrac{x}{5}$

17. $z + 30$ **18.** $2x - \dfrac{y}{3}$ **19.** (a) and (d) only **20.** 60 cm^2 **21a.** 3 less than twice a number, x, or the difference of twice a number, x, and 3 **21b.** Twice the difference of a number, x, and 3

Chapter 1 Answers
Section 1.1

Five-Minute Warm-Up **1.** $\dfrac{1}{2}$ **2.** $\dfrac{1}{6}$ **3.** x **4.** 60 **5.** $-6x + 15$ **6.** -1 **7.** $x + 7$ **8.** $7x$ **9.** -19 **10.** yes

Guided Practice **1.** Substitute the value into the original equation. If this results in a true statement, then the value for the variable is a solution to the equation. **2a.** Remove any parentheses using the Distributive Property. **2b.** Combine like terms on each side of the equation. **2c.** Use the Addition Property of Equality to get all variable terms on one side of the equation and all constants on the other side. **2d.** Use the Multiplication Property of Equality to get the coefficient of the variable to be 1. **2e.** Check your answer to be sure that is satisfies the original equation. **3a.** 12 **3b.** 12; 12 **3c.** $12\left(\dfrac{2x + 3}{2}\right) - 12\left(\dfrac{x - 1}{4}\right) = \left(\dfrac{x + 5}{12}\right) \cdot 12$

3d. $6(2x + 3) - 3(x - 1) = (x + 5) \cdot 1$ **3e.** the Distributive Property **3f.** $9x + 21 = x + 5$
3g. $9x + 21 - x = x + 5 - x$ **3h.** $8x + 21 = 5$ **3i.** $8x + 21 - 21 = 5 - 21$ **3j.** $8x = -16$ **3k.** $x = -2$
3l. $\{-2\}$ **4a.** false **4b.** $\varnothing$ **4c.** contradiction **5a.** true **5b.** $\{x \mid x \text{ is any real number}\}$ **5c.** identity

Do the Math 1. yes 2. yes 3. $\{-3\}$ 4. $\{1\}$ 5. $\{5\}$ 6. $\{-1\}$ 7. $\varnothing$ 8. $\left\{\dfrac{6}{5}\right\}$ 9. $\left\{-\dfrac{7}{3}\right\}$ 10. $\{-3\}$

11. $\left\{\dfrac{17}{2}\right\}$ 12. $\{x \mid x \text{ is any real number}\}$ 13. $\{-20\}$ 14. $\varnothing$ 15. $a = 2$ 16. $a = 5$ 17. $x = -\dfrac{8}{5}$

18. $x = \left\{\dfrac{5}{2}\right\}$ 19. We "solve" an equation, which means to find the number or numbers that will replace the variable so that the left side of the equation will be equal to the right side of the equation. We "simplify" an expression (using the Distributive Property or by combining like terms) to form an equivalent algebraic expression.

Section 1.2

Five-Minute Warm-Up 1. Possible answers: sum, plus, more than, exceeds by, in excess of, added to, increased by 2. Possible answers: difference, minus, subtracted from, less, less than, decreased by 3. Possible answers: product, times, of, twice 4. Possible answers: quotient, divided by, per, ratio. 5. $2x + 3$

6. $5(y + 1)$ 7. $5z + 20$ 8. $\dfrac{2n}{3}$ 9. $\dfrac{x}{2} - 12$ 10. $x - 15$ 11. 0.05 12. 7.5%

Guided Practice 1. Direct Translation: problems where we translate from English into Mathematics by using key words in the verbal description; Mixture: problems where two or more quantities are combined; Geometry: problems where the unknown quantities are related through geometric formulas; Uniform Motion: problems where an object travels at a constant speed; Work Problems: problems where two or more entities join forces to complete a job. 2. Identify what you are looking for; Give names to the unknowns; Translate the problem into the language of mathematics; Solve the equation(s); Check the reasonableness of your answer; Answer the question 3a. $n + 1$ 3b. $n + 2$ 4a. $x + 2$ 4b. $x + 4$ 4c. $x + 6$ 5a. $p + 2$ 5b. $p + 4$ 5c. $p + 6$ 6. $I = Prt$; I = Interest, P = Principal, r = interest rate expressed as a decimal, t = time

7a. $11 - p$ 7b. Chocolates: $4.50, p, 4.5p$; Truffles: $7.50, 11 - p, 7.5(11 - p)$; Blend: no value, 11, 58.50

7c. $4.5p + 7.5(11 - p) = 58.5$; $p = 8$ 7d. Andy bought 8 pounds of chocolates and 3 pounds of truffles 8a. The distances add to 63 miles. 8b. Slower boat: $r, 4.5$ hours, $4.5r$; Faster boat: $r + 6, 4.5$ hours, $4.5(r + 6)$ Total: no value, no value, 63 miles 8c. $4.5r + 4.5(r + 6) = 63$; $r = 4$ 8d. Slower boat is traveling at 4 mph; Faster boat is traveling at 10 mph

Do the Math 1. $\dfrac{500}{9}$ 2. 75% 3. $10 - z = 6$; $z = 4$ 4. $2y + 3 = 16$; $y = \dfrac{13}{2}$ 5. $5x = 3x - 10$; $x = -5$

6. $0.4x = x - 10$; $x = \dfrac{50}{3}$ 7. 24 and 32 8. 23, 25, 27, and 29 9. $3750 in savings; $6250 in stock

10. 6 nickels; 42 dimes 11. $2.5\,l$ 12. 20 lb of almonds; 30 lb of peanuts 13. 3.4 hours

Section 1.3

Five-Minute Warm-Up 1a. $A = s^2$; $P = 4s$ 1b. $A = lw$; $P = 2l + 2w$ 1c. $A = \dfrac{1}{2}bh$; $P = a + b + c$

1d. $A = \dfrac{1}{2}h(B + b)$; $P = a + b + c + B$ 1e. $A = ah$; $P = 2a + 2b$ 1f. $A = \pi r^2$; $C = 2\pi r$ or πd

2a. 15.961 2b. 15.961 3a. -0.10 3b. -0.09 4a. 10 4b. 9 5a. 100.7 5b. 100.7 6a. $5.77 6b. $5.76
Guided Practice 1. Get the variable by itself, with a coefficient of 1, on one side of the equation and all of the constants and any other variables on the other side of the equation. 2a. $h = \dfrac{3V}{B}$ 2b. $P = \dfrac{A}{1 + rt}$

3a. the sum of the measures of the angles of a triangle equals 180 degrees or $x° + y° + z° = 180$

3b. $x - 15; \dfrac{x}{2} + 45$ 3c. $x - 15 + x + \dfrac{x}{2} + 45 = 180$ 3d. $x = 60$ 3e. yes 3f. $45°; 60°; 75°$

4a. $w = \dfrac{P - 2l}{2}$ 4b. width = 6.5 cm 4c. Answers will vary

Do the Math 1. $k = \dfrac{y}{x}$ 2. $m = \dfrac{y - b}{x}$ 3. $C = \dfrac{5}{9}(F - 32)$ 4. $h = \dfrac{2A}{b}$ 5. $S = \dfrac{a(1 - r)^n}{1 - r}$

Do the Math 6. $b = \dfrac{2A - Bh}{h}$ or $\dfrac{2A}{h} - B$ **7.** $y = -2x + 10$ **8.** $y = \dfrac{4}{15}x - 2$ **9.** $y = 4x - 5$ **10.**

$y = -\dfrac{6}{5}x + 24$ **11a.** $A = \dfrac{217 - M}{0.85}$ **11b.** 67 years old **12.** 60°, 120° **13.** 40°, 50° **14.** 32°, 96°, 52°

15a. 27.53 ft^2 **15b.** 23.71 ft **15c.** using part (a) \$227.12; using π key and not rounding \$227.16

Section 1.4

Five-Minute Warm-Up 1. < **2.** > **3.** > **4.** = **5.** > **6.** < **7a.** > or < **7b.** ≥ or ≤ **7c.** $\dfrac{3}{8} < x < \dfrac{1}{2}$

Guided Practice 1. none **2.** TRUE **3.** TRUE **4.** FALSE **5a.** $-2; -2$ **5b.** $5x > -15$ **5c.** $\dfrac{5x}{5} > \dfrac{-15}{5}$

5d. $x > -3$ **5e.** $\{x \mid x > -3\}$ **5f.** $(-3, \infty)$ **5g.** See Graphing Answer Section **6a.** $-7; -7$ **6b.** $3x \geq 7x - 8$

6c. $-7x; -7x$ **6d.** $-4x \geq -8$ **6e.** $x \leq 2$ **6f.** $\{x \mid x \leq 2\}$ **6g.** $(-\infty, 2]$ **6h.** See Graphing Answer Section

7a. ≥ **7b.** ≥ **7c.** > **7d.** > **7e.** ≤ **7f.** ≤ **7g.** < **7h.** < **8a.** at least **8b.** $25{,}000 + 0.20v \geq 36{,}000$

8c. $v \geq 55{,}000$ **8d.** yes **8e.** Nghiep must have annual sales of at least \$55,000 in order to earn \$36,000.

Do the Math 1 – 4. See Graphing Answer Section **1.** $\{x \mid x < 3\}$ **2.** $\{x \mid x \leq 4\}$ **3.** $\{x \mid x > -3\}$

4. $\{x \mid x \geq -4\}$ **5.** $(-\infty, 6]$ **6.** $\left(-\infty, -\dfrac{5}{2}\right]$ **7.** $(3.2, \infty)$ **8.** $\left(\dfrac{7}{2}, \infty\right)$ **9.** $\left(-\infty, -\dfrac{8}{5}\right)$ **10.** $\left(-\infty, -\dfrac{2}{5}\right)$

11. $\left(-\infty, \dfrac{1}{2}\right)$ **12.** $[-3, \infty)$ **13.** $(-\infty, 3)$ **14.** $(5, \infty)$ **15.** at least 91.5 **16.** 132 miles **17.** As written, this

notation says that $x > 7$ and at the same time $x > 4$, which simplifies to x > 7.

Section 1.5

Five-Minute Warm-Up 1. See Graphing Answer Section **2.** (b) only **3a.** -14 **3b.** 6 **3c.** 4

4. $y = \dfrac{4}{3}x + 4$ **5a.** 12 **5b.** 0 **5c.** 125

Guided Practice 1. See Graphing Answer Section **2.** origin **3.** left **4.** 2 **5a.** $-2; -7; (-2, -7)$

5b. $-1; -4; (-1, -4)$ **5c.** $0; -1; (0, -1)$ **5d.** $1; 2; (1, 2)$ **5e.** $2; 5; (2, 5)$ **6.** $(a, 0)$ and $(b, 0)$ **7.** $(0, c)$ **8.**

If one unit is manufactured and sold, there will be a gain (profit) of \$50. **9.** 2 sec after the ball leaves the

hand of the thrower, it will be 8 ft above the ground.

Do the Math 1A. $(4, -3)$; IV **1B.** $(-3, -1)$; III **1C.** $(-2, 1)$; II **2.** See Graphing Answer Section

2D. y-axis **2E.** 1 **2F.** x-axis **3.** (b) only **4.** (a) and (d) **5.** $(2, 0)$; $(0, -4)$ **8.** **6 – 9.** See Graphing Answer

Section **10.** $a = -\dfrac{7}{3}$ **11.** $b = -13$

Section 1.6

Five-Minute Warm-Up 1. $\{6\}$ **2.** $\left\{-\dfrac{2}{5}\right\}$ **3.** $y = \dfrac{5}{2}x + 10$ **4.** $y = \dfrac{4}{3}x - 2$ **5.** -2 **6.** 2 **7.** $-3x + 15$

8. $\dfrac{5}{4}x - 10$

Guided Practice 1. standard **2.** $y = 0$ **3.** $x = 0$ **4.** slope **5.** $m = \dfrac{y_2 - y_1}{x_2 - x_1}$ **6a.** slants upward from left

to right **6b.** slants downward from left to right **6c.** horizontal **6d.** vertical **7.** $y - y_1 = m(x - x_1)$

8a. $-2; 4; -1$ **8b.** $y - y_1 = m(x - x_1)$ **8c.** $y - (-1) = -2(x - 4)$ **8d.** $y = -2x + 7$ **9a.** $m = \dfrac{y_2 - y_1}{x_2 - x_1}$

9b. $x_1 = 2, y_1 = -2, x_2 = -2, y_2 = 6$ **9c.** $m = \dfrac{6 - (-2)}{-2 - 2} = \dfrac{8}{-4} = -2$ **9d.** $y - y_1 = m(x - x_1)$

9e. $m = -2, x_1 = 2, y_1 = -2$ **9f.** $y - (-2) = -2(x - 2)$ **9g.** $y + 2 = -2x + 4$ **9h.** 2;

$y + 2 - 2 = -2x + 4 - 2$ **9i.** $y = -2x + 2$; yes **9j.** no **9k.** -2 **9l.** $(0, 2)$ **9m.** See Graphing Answer Section

Do the Math 1 – 4. See Graphing Answer Section **5.** $-\dfrac{12}{5}$ **6.** undefined **7.** See Graphing Answer

Section **8.** $y = -x$ **9.** $y = 4x - 9$ **10.** $y = -\dfrac{4}{3}x - \dfrac{5}{3}$ **11.** $y = -\dfrac{1}{3}x - \dfrac{2}{3}$ **12.** $m = 3$; y-intercept: $(0,2)$

13. $m = -2$; y-intercept is $(0,3)$ **14.** $m = $ undefined; no y-intercept **15.** $m = 0$; y-intercept is $(0,-4)$

16. $g(x) = 3x + 2; -7$ **17.** $F(x) = -\dfrac{4}{5}x + \dfrac{33}{5}; \dfrac{39}{5}$ **18.** Vertical line of the form $x = a$, where $a \neq 0$

19. Horizontal line of the form $y = b$, where $b \neq 0$

Section 1.7

Five-Minute Warm-Up **1.** 1 **2.** $-\dfrac{1}{5}$ **3.** $-\dfrac{2}{3}$ **4.** 3 **5.** $y = \dfrac{7x + 9}{3}$ or $y = \dfrac{7}{3}x + 3$ **6.** -4 **7.** undefined

8. 0

Guided Practice **1.** Two lines are parallel if they never intersect. **2.** slope; y-intercept **3.** $=$; $\neq$

4a. $y = \dfrac{3}{4}x - \dfrac{1}{2}$ **4b.** $\dfrac{3}{4}$ **4c.** $\dfrac{3}{4}$ **4d.** $\dfrac{3}{4}; 4; -1$ **4e.** $y - y_1 = m(x - x_1)$ **4f.** $y - (-1) = \dfrac{3}{4}(x - 4)$

4g. $y = \dfrac{3}{4}x - 4$ **4h.** See Graphing Answer Section **5.** Two lines are perpendicular if they intersect at right

angles $(90°)$. **6.** -1 **7.** $-1; -\dfrac{1}{m_2}$ **8a.** $y = 3x - 6$ **8b.** 3 **8c.** $-\dfrac{1}{3}$ **8d.** $-\dfrac{1}{3}; -9; 1$

8e. $y - y_1 = m(x - x_1)$ **8f.** $y - 1 = -\dfrac{1}{3}(x - (-9))$ **8g.** $y = -\dfrac{1}{3}x - 2$ **8h.** See Graphing Answer Section

Do the Math **1a.** $-10; -\dfrac{1}{10}$ **1b.** $2; \dfrac{-1}{2}$ **1c.** $\dfrac{3}{2}; \dfrac{3}{2}$ **1d.** 0; undefined **1e.** undefined; undefined

2. perpendicular **3.** neither **4.** perpendicular **5.** $y = -3x + 11$ **6.** $y = -\dfrac{1}{4}x + 2$ **7.** $x = 2$

8. $y = -\dfrac{5}{2}x + 2$ **9.** $y = -2x - 5$ **10.** $x = 2$ **11.** $y = \dfrac{1}{3}x - 2$ **12.** $y = -\dfrac{1}{4}x + \dfrac{1}{4}$ **13.** $B = -4$

14. for $\overline{AB}, \overline{CD}$ $m = \dfrac{1}{3}$; for $\overline{BC}, \overline{DA}$ $m = 4$; Yes, it is a parallelogram.

Section 1.8

Five-Minute Warm-Up **1.** yes **2.** no **3.** $\{x | x \geq 5\}$ **4.** $\{x | x \leq 0\}$ **5.** $\{n | n < 12\}$ **6.** $\{a | a > -12\}$

Guided Practice **1.** True **2a.** dashed **2b.** solid **3.** half planes **4.** shade **5a.** $3x - 4y = 12$ **5b.** $(4,0)$

5c. $(0,-3)$ **5d.** dashed **5e.** Answers may vary; $(0, 0)$ **5f.** false **5g.** opposite half plane **5h.** See Graphing

Answer Section **6a.** $45x + 65y \geq 4000$ **6b.** yes **6c.** no

Do the Math **1.** (a) and (b) **2.** (a) only **3 – 8.** See Graphing Answer Section **9.** $x \leq y + 12$

10. $x + y \geq -3$ **11a.** $0.10g + 0.25s \leq 3.00$ **11b.** no **11c.** yes **12a.** $0.50x + 2y \geq 1000$ **12b.** no

12c. yes **13.** $y > \dfrac{4}{3}x - \dfrac{5}{3}$

Chapter 2 Answers
Section 2.1

Five-Minute Warm-Up **1a.** $(-1, 0]$ **1b.** $[4, 7)$ **2a.** $(-\infty, -4]$ **2b.** $(-2, \infty)$ **3 – 5.** See Graphing

Answer Section

Guided Practice **1.** A relation is a mapping that pairs elements of one set with elements of a second set.

2. inputs; x **3.** outputs; y **4.** mapping; a set of ordered pairs; graph; equation **5a.** $(-\infty, \infty)$ **5b.** $[-3, \infty)$

6a. $[-3, 3]$ **6b.** $[-4, 4]$ **7.** (1, -2), (-2, -1), (-3, 0), (-2, 1), (1, 2) See Graphing Answer Section **7a.** $[-3, \infty)$

7b. $(-\infty, \infty)$

Do the Math **1.** map: $(30, 7.09)$, $(35, 7.09)$, $(40, 8.40)$, $(45, 11.29)$; domain: $\{30, 35, 40, 45\}$;
range: $\{7.09, 8.40, 11.29\}$ **2.** domain: $\{-2, -1, 0, 1, 2\}$; range: $\{-3, 0, 3, 6\}$ **3.** domain: $\{-2, -1, 0, 1, 2\}$;
range: $\{-8, -1, 0, 1, 8\}$ **4.** domain: $\{-3, 0, 3\}$; range: $\{-3, 0, 3\}$ **5.** domain: $[-3, 3]$; range: $[-2, 2]$
6. domain: $\{-2, 0, 1, 2\}$; range: $\{-2, -1, 0, 3\}$ **7.** domain: $(-\infty, \infty)$; range: $[-3, \infty)$
8. domain: $(-\infty, \infty)$; range: $(-\infty, \infty)$ **9.** domain: $(-\infty, \infty)$; range: $(-\infty, \infty)$ **10.** domain: $(-\infty, \infty)$;
range: $[-2, \infty)$ **11.** domain: $(-\infty, \infty)$; range: $(-\infty, 8]$ **12.** domain: $(-\infty, \infty)$; range: $[-2, \infty)$
13. domain: $[2, \infty)$; range: $(-\infty, \infty)$ **14.** domain: $(-\infty, \infty)$; range: $(-\infty, \infty)$

Section 2.2

Five-Minute Warm-Up **1.** 17 **2.** -6 **3.** $(-16, \infty)$ **4.** $\{x \mid x \le 100\}$ **5a.** domain: $\{-2, -4, -6, -8\}$

5b. range: $\{1, 3, 5, 7\}$ **6.** $\dfrac{1}{5}$ **7.** -5 **8.** 3 **9.** 108.0 in.^{3}

Guided Practice **1.** A function is a relation in which each element of the domain (inputs) corresponds to
exactly one element in the range (outputs). **2a.** yes; domain: $\{3, 4, 5\}$; range: $\{-6, -1\}$ **2b.** no
3a. $2x; -2x$ **3b.** no **4a.** yes **4b.** no **5.** A graph represents a function if and only if every possible vertical
line intersects the graph in at most one point. **6.** (c) and (d) only **7.** independent; domain **8.** dependent;
range **9.** 52 **10.** $-6a + 2$ **11.** 4 **12.** $-3k + 1$ **13a.** N, the number of trucks produced **13b.** t, time worked in
hours **13c.** 300; After 5 hours, the factory produced 300 trucks.
Do the Math **1.** No **2.** No **3.** Yes **4.** Yes **5.** Yes **6.** Yes **7.** No **8.** Yes **9.** No **10.** No
11. Yes **12a.** 10 **12b.** -5 **12c.** $6x + 1$ **12d.** $3x + 7$ **13a.** -9 **13b.** 1 **13c.** $-4x - 3$ **13d.** $-2x - 7$
14. -20 **15.** 3 **16.** $C = 3$ **17.** $B = 5$

Section 2.3

Five-Minute Warm-Up **1.** $\left\{\dfrac{1}{2}\right\}$ **2.** $\{6\}$ **3 – 4.** See Graphing Answer Section **5a.** domain: $[-4, 4]$

5b. range: $[-1, 2]$ **6a.** domain: $[-3, 2]$ **6b.** range: $\left[-\dfrac{7}{6}, \dfrac{10}{3}\right]$

Guided Practice **1.** zero **2.** $\left\{t \mid t \ne -\dfrac{3}{2}\right\}$ **3.** $(-\infty, \infty)$ $(-2, 8)(-1, 6)(0, 4)(1, 2)(2, 0)(3, 2)(4, 4)$

4. See Graphing Answer Section **5a.** $(-\infty, \infty)$ **5b.** $[-2, \infty)$ **5c.** x-intercepts: $(-2, 0)$, $(2, 0)$; y-intercept:
$(0, -2)$ **6a.** yes **6b.** $-7; (6, -7)$ **6c.** $2; (2, 3)$ **7.** x; zero **8.** $(2, 0)$, $(-2, 0)$

Do the Math **1.** $(-\infty, \infty)$ **2.** $\left(-\infty, -\dfrac{1}{2}\right) \cup \left(-\dfrac{1}{2}, \infty\right)$ **3.** $(-\infty, \infty)$ **4.** $\left(-\infty, -\dfrac{5}{6}\right) \cup \left(-\dfrac{5}{6}, \infty\right)$

5. $(-\infty, \infty)$ **6 – 7.** See Graphing Answer Section **8a.** $(-\infty, \infty)$ **8b.** $(-\infty, \infty)$ **8c.** x-intercept: $(3, 0)$;
y-intercept: $(0, -1)$ **8d.** $x = 3$ **9a.** $(-\infty, \infty)$ **9b.** $(-\infty, \infty)$ **9c.** x-intercepts: $(-2, 0)(1, 0)(4, 0)$;
y-intercept: $(0, 2)$ **9d.** $x = -2, x = 1, x = 4$ **10a.** 8 **10b.** 0, 7 **10c.** $(-4, 10)$ **10d.** $(0, 5)$ **10e.** yes; $x = -4$
11a. no **11b.** $17; (4, 17)$ **11c.** $-3; (-3, -4)$ **11d.** no **12a.** $[0, \infty)$ **12b.** 381.7 cm^3

Section 2.4

Five-Minute Warm-Up **1– 2.** See Graphing Answer Section **3.** $y = -3x - 1$ **4.** $\{-0.9\}$ **5.** $\left\{y \mid y \ge -\dfrac{1}{24}\right\}$

Guided Practice **1.** linear; line **2.** slope; y-intercept **3a.** $(0, -2)$ **3b.** $\dfrac{3}{2}$ **3c.** See Graphing Answer

Section **4.** $mx + b = 0$ **5a.** $(0, \infty)$ **5b.** 66 ft **5c.** 15 ft **5d.** when the width exceeds 3 ft
6. fixed; a **7a.** \$4500 **7b.** negative **7c.** $V(x) = 22,500 - 4500x$ **7d.** $[0, 5]$ **7e.** \$4500 **7f.** after 2.75
years **7g.** time, x **7h.** the value, V **8.** 2 **9a.** slope **9b.** $y - y_1 = m(x - x_1)$

Do the Math 1 – 4. See Graphing Answer Section 5. $x = 8$ 6. $x = 4$ 7a. $\{-4\}$ 7b. $-\dfrac{1}{3}$

7c. $-\dfrac{1}{3}$ 7d. $\{x\,|\,x > -4\}$ 8. See Graphing Answer Section; $\left(-4, -\dfrac{1}{3}\right)$ 9a. $g(x) = 3x + 2$ 9b. -7

10a. age, a 10b. Birth rate, B 10c. $[15, 44]$ 10d. 23.5 per 1000 women 10e. 37 years old

Section 2.5

Five-Minute Warm-Up 1a. $\{x\,|-4 \le x < -1\}$ 1b. $[-4, -1)$ 2. See Graphing Answer Section

3. $(-3, 5]$ 4. $\{-3\}$ 5. $\{x\,|\,x \le -6\}$ 6. $\{x\,|\,x > -2\}$

Guided Practice 1. $A \cap B$; and 2. $A \cup B$; or 3a – 3b. See Graphing Answer Section 4a. $5x < 10$

4b. $x < 2$ 4c. $x \ge -4$ 4d – 4f. See Graphing Answer Section 5. $a < x < b$

6a. $-8 + 3 \le 5x - 3 + 3 \le 7 + 3$ 6b. $-5 \le 5x \le 10$ 6c. $-1 \le x \le 2$ 6d. See Graphing Answer Section

7a. $-\dfrac{3}{2}x > 3$ 7b. $x < -2$ 7c. $7x > 14$ 7d. $x > 2$ 7e – 7g. See Graphing Answer Section

7h. $(-\infty, -2) \cup (2, \infty)$

Do the Math 1. $\{2, 3, 4, 5, 6, 7, 8, 9\}$ 2. $\{4, 6\}$ 3. $\varnothing$ 4. $[-2, 2]$ 5. $(-\infty, \infty)$

6. $(0, 5]$ 7. $(-\infty, 0) \cup [6, \infty)$ 8. $x = 1$ 9. $(-\infty, 2) \cup [5, \infty)$ 10. $\left(-2, \dfrac{4}{7}\right]$ 11. $(-\infty, -2] \cup (3, \infty)$

12. $(-\infty, -10) \cup (1, \infty)$ 13. $(2, 4]$ 14. $\left(-\dfrac{5}{3}, \dfrac{11}{4}\right]$ 15. $(-4, 1)$ 16. $\left[-\dfrac{9}{4}, 3\right)$ 17. $\left[-\dfrac{7}{3}, 3\right)$

18. $a = -15; b = -1$ 19. $60 < x < 90$ 20. approximately 756.7 kwh up to approximately 1953.1 kwh

Section 2.6

Five-Minute Warm-Up 1a. 12 1b. 0 1c. $\dfrac{3}{4}$ 1d. 5.2 2. $|45|$ 3. $|-12|$ 4a. $\{4\}$ 4b. $\{7\}$

5a. $\{x\,|\,x > -6\}$ 5b. $\{x\,|\,x \le 3\}$

Guided Practice 1. $u = a; u = -a$ 2. isolate the absolute value expression 3a. $|3x - 1| = 2$

3b. $3x - 1 = 2$ or $3x - 1 = -2$ 3c. $1; -\dfrac{1}{3}$ 3d. $\left\{1, -\dfrac{1}{3}\right\}$ 4. $u = v; u = -v$ 5. $-a < u < a$; $-a \le u \le a$

6a. $-9 \le 4x - 3 \le 9$ 6b. $-6 \le 4x \le 12$ 6c. $-\dfrac{3}{2} \le x \le 3$ 6d. See Graphing Answer Section 6e. $\left[-\dfrac{3}{2}, 3\right]$

7. $u > a$ or $u < -a$; $u \ge a$ or $u \le -a$ 8a. $|8x + 3| > 3$ 8b. $8x + 3 > 3$ or $8x + 3 < -3$

8c. $x > 0$ or $x < -\dfrac{3}{4}$ 8d. See Graphing Answer Section 8e. $\left(-\infty, -\dfrac{3}{4}\right) \cup (0, \infty)$

Do the Math 1. $\left\{-\dfrac{7}{2}, \dfrac{13}{2}\right\}$ 2. $\{0, 8\}$ 3. $\{y\,|-10 < y < 2\}$ 4. $\left\{x\,|-1 \le x \le \dfrac{7}{3}\right\}$

5. $\{x\,|\,x \le -11 \text{ or } x \ge 3\}$ 6. $\{z\,|\,z \text{ is any real number}\}$ 7. $\varnothing$ 8. $\left(-\dfrac{4}{5}, 2\right)$ 9. $\varnothing$ 10. $(-\infty, \infty)$

11. $\left\{-1, -\dfrac{1}{2}\right\}$ 12. $\left(-\infty, \dfrac{1}{2}\right) \cup (1, \infty)$ 13. $\varnothing$ 14. $\{y\,|\,y \text{ is any real number}\}$ 15. $|x - (-4)| < 2$

16. $|2x - 7| > 3$ 17. <234.64 days or >297.36 days 18. The absolute value of every real is number is positive or 0 and therefore greater than any negative number.

Section 3.1

Five-Minute Warm-Up **1.** 17 **2.** Yes **3.** See Graphing Answer Section **4.** $y = x + 1$

5. slope: $\dfrac{1}{3}$; y-intercept: $(0, 3)$ **6.** -15 **7.** $\{1\}$

Guided Practice **1a.** inconsistent; the lines are parallel **1b.** consistent and dependent; the lines coincide **1c.** consistent and independent; the lines intersect **2a.** $-y = -3x - 14$ **2b.** $y = 3x + 14$ **2c.** $5x + 2y = -5$ **2d.** $5x + 2(3x + 14) = -5$ **2e.** $5x + 6x + 28 = -5$ **2f.** $11x + 28 = -5$ **2g.** $11x = -33$ **2h.** $x = -3$ **2i.** 5 **2j.** $(-3, 5)$ **3a.** $(1) -10x - 5y = 20$; $(2) 3x + 5y = 29$ **3b.** $-7x = 49$ **3c.** $x = -7$ **3d.** $2(-7) + y = -4$ **3e.** $y = 10$ **3f.** $(-7, 10)$ **4.** The variables will both be eliminated and a false statement will remain. **5.** The variables will both be eliminated and a true statement will remain **6.** $\{(x, y) \mid -x + 3y = 1\}$ or $\{(x, y) \mid 2x - 6y = -2\}$

Do the Math **1a.** no **1b.** yes **2a.** no **2b.** no **3.** $(2, 0)$ **4.** $(1, -4)$ **5.** $(4, 2)$ **6.** $\left(\dfrac{5}{3}, -\dfrac{5}{2}\right)$

7. $\varnothing$ **8.** $\left(\dfrac{2}{9}, -\dfrac{13}{9}\right)$ **9.** $(5, -11)$ **10.** $\left\{(x, y) \mid y = \dfrac{1}{2}x + 2\right\}$ or $\{(x, y) \mid x - 2y = -4\}$ **11.** $\varnothing$

12. no solution **13.** exactly one solution **14a.** $y = -\dfrac{1}{2}x + \dfrac{5}{2}$ **14b.** $y = 2x$ **14c.** $(1, 2)$

Section 3.2

Five-Minute Warm-Up **1.** $x + 12 = 5$ **2.** $y - 6 = 2y$ **3.** $2(l - 6) = w$ **4.** $2h - 10 = \dfrac{2}{w}$ **5.** $12000 - s$

6. $\$37.50$ **7.** $C(x) = 42x + 4000$

Guided Practice **1a.** The number of child's tickets. **1b.** $(1)\ a + c = 13$; $(2)\ 26a + 18.50c = 278$ **1c.** $(5, 8)$ **1d.** Yes; Yes **1e.** 5 adult and 8 child's tickets were purchased. **2a.** $P = 2l + 2w$ **2b.** $l = w + 10$ **2c.** $(1)\ 2l + 2w = 125$; $l - w = 10$ **2d.** $(36.25, 26.25)$ **2e.** The dimensions are 36.25 ft by 26.25 ft **3a.** Peanuts: p; 2; $2p$; Trail Mix: t, 5, $5t$; Blend 10, 3.20, 32.00 **3b.** $(1)\ p + t = 10$; $2p + 5t = 32$ **3c.** $(6, 4)$; The backpacker used 6 pounds of peanuts and 4 pounds of trail mix. **4a.** $r - c$ **4b.** $r + c$ **5a.** With the wind: 36, $r + c$, 3; Against the wind: 32, $r - c$, 4 **5b.** $(1)\ 3r + 3c = 36$; $4r - 4c = 32$ **5c.** $(10, 2)$; The speed of the cyclist is 10 mph and the speed of the wind is 2 mph. **6.** The point(s) of intersection of the two graphs. **7.** revenue = costs

Do the Math **1.** 14 and 11 **2.** 15 and 8 **3.** 25 dimes **4.** 72.5 cm by 57.5 cm **5.** $\$32,000$ in bonds **6.** 10 lb. of chocolate covered peanuts and 40 lb. of chocolate covered almonds **7.** 25 units of powder A and 10 units of powder B **8.** 150 mph; 20 mph **9.** 0.3 hour or 18 minutes

Section 3.3

Five-Minute Warm-Up **1.** -10 **2.** $(-5, -15)$ **3.** $\varnothing$ **4.** $\{(x, y) \mid 4x + y = 3\}$ or $\{(x, y) \mid 8x + 2y = 6\}$

Guided Practice **1.** a plane **2a.** consistent; independent **2b.** inconsistent **2c.** consistent; dependent **3a.** $-3z - 9y - 9z = -27$ **3b.** $3x + 5y + 4z = 8$ **3c.** $-4y - 5z = -19$ **3d.** $-5x - 15y - 15z = -45$ **3e.** $5x + 3y + 7z = 9$ **3f.** $-12y - 8z = -36$ **3g.** $12y + 15z = 57$ **3h.** $-12y - 8z = -36$ **3i.** $7z = 21$ **3j.** $z = 3$ **3k.** $y = 1$ **3l.** $x = -3$ **3m.** $(-3, 1, 3)$ **4.** inconsistent; $\varnothing$ **5.** consistent and dependent **6a.** $x = 5z - 3$ **6b.** $y = 10z - 8$ **6c.** $x = 5z - 3$, $y = 10z - 8$ **7a.** $a + b + c = 600$

7b. $80a + 60b + 25c = 33,500$ **7c.** $80a + \dfrac{3}{5}b(60) + \dfrac{4}{5}c(25) = 24,640$

Do the Math **1a.** Yes **1b.** Yes **2.** $(-2, 6, 6)$ **3.** $\left(\dfrac{13}{2}, 0, -\dfrac{3}{2}\right)$ **4.** $(0, -3, 1)$ **5.** $\varnothing$

6. $\left(2, \dfrac{3}{2}, \dfrac{1}{2}\right)$ **7a.** $a + b + c = 2$; $4a + 2b + c = 9$ **7b.** $a = 3$; $b = -2$; $c = 1$ **7c.** $f(x) = 3x^2 - 2x + 1$

8. Potato: 1.5; Chicken: 1.5; Coke: 1

Section 3.4

Five-Minute Warm-Up **1.** $-2, 1, -3$ **2.** 12 **3.** $y = \dfrac{7}{5}x - 2$ **4.** $x = -\dfrac{4}{9}y - \dfrac{4}{3}$ **5.** $-6x + 27y - 3z$

6. $-10x + 5y - 15z$ **7a.** -29 **7b.** 13

Guided Practice **1.** 2×4 **2.** standard; 0 **3.** See Graphing Answer Section **4.** $\begin{cases} x + y = 2 \\ -3x + y = 10 \end{cases}$

5a. Interchange any 2 rows. **5b.** Replace a row by a non-zero multiple of that row. **5c.** Replace a row by the sum of that row and a non-zero multiple of another row. **6a – 6b.** See Graphing Answer Section **7.** When there are ones on the main diagonal (when the row and column number are the same) and zeros below the ones. This is also called triangular form. **8a – 8d.** See Graphing Answer Section **8e.** $(1, 2, -2)$

9. $\{(x, y) | x + 4y = 2\}$ **10.** $\varnothing$

Do the Math **1 – 2b.** See Graphing Answer Section **3.** $\{(x, y) | 5x - 2y = 3\}$ **4.** $\left(\dfrac{4}{3}, -\dfrac{5}{3}\right)$

5. $\varnothing$ **6.** $(0, -5, 4)$ **7.** \$7000 in savings; \$3000 in Treasury bonds; \$2000 in the mutual fund

Section 3.5

Five-Minute Warm-Up **1.** -1 **2a.** undefined **2b.** 0 **3.** $\dfrac{6}{5}$ **4.** $-\dfrac{31}{20}$ **5.** $\{-3\}$ **6.** $\{28\}$ **7.** $\varnothing$

Guided Practice **1a.** 14 **1b.** -25 **2a.** coefficient **2b.** the constants **2c.** second; y **3.** $\begin{cases} 3x - 5y = 9 \\ x + 2y = 2 \end{cases}$

4a. $\begin{vmatrix} 3 & -5 \\ 1 & 2 \end{vmatrix}$ **4b.** $\begin{vmatrix} 9 & -5 \\ 2 & 2 \end{vmatrix}$ **4c.** $\begin{vmatrix} 3 & 9 \\ 1 & 2 \end{vmatrix}$ **5a.** $x = \dfrac{D_x}{D}$ **5b.** $y = \dfrac{D_y}{D}$ **6.** See Graphing Answer Section.

7. See the determinants in the Graphing Answer Section **7b.** 22 **7c.** -66 **7d.** 11 **7e.** 22 **7f.** $\dfrac{-66}{22} = -3$

7g. $\dfrac{D_y}{D} = \dfrac{11}{22} = \dfrac{1}{2}$ **7h.** $\dfrac{D_z}{D} = \dfrac{22}{22} = 1$ **7i.** $\left(-3, \dfrac{1}{2}, 1\right)$ **8.** inconsistent; $\varnothing$ **9.** consistent; dependent; infinitely many

Do the Math **1.** 14 **2.** -4 **3.** -5 **4.** $(5, -4)$ **5.** $\left(\dfrac{7}{3}, \dfrac{5}{6}\right)$ **6.** $(3, 2, -1)$ **7.** Cramer's Rule does not

apply **8.** $x = -3$ **9.** area is 12.5 sq units

Section 3.6

Five-Minute Warm-Up **1.** Yes **2.** No **3.** $\{x | x < -3\}$ **4.** See Graphing Answer Section

Guided Practice **1.** graphing **2.** dashed **3.** solid **4.** the opposite half-plane **5.** intersection
6. See Graphing Answer Section **7a.** $5x + 8y \le 40$ **7b.** $y \ge 2x$ **7c.** $x \ge 0$; $y \ge 0$

7d. See Graphing Answer Section **7e.** $(0, 5)\,(0, 0), \left(\dfrac{40}{21}, \dfrac{80}{21}\right)$

Do the Math **1.** (b) **2.** (b) **3-7.** See Graphing Answer Section **7.** The graph is unbounded. Corner points:

$(0, 8)$; $(6, 2)$; $(12, 0)$ **8a.** See Graphing Answer Section **8b.** $\left(0, \dfrac{800}{3}\right)$; $(0, 0)$; $\left(\dfrac{400}{3}, 0\right)$

Chapter 4 Answers

Section 4.1

Five-Minute Warm-Up   **1.** -1 **2.** 5 **3.** $4x^2 + 2x$ **4.** $-48x - 16$ **5.** $5p + 12$ **6.** xy^2 **7.** 27 **8.** -29

Guided Practice   **1a.** A single term which is the product of a constant and a variable with whole number exponents.   **1b.** The constant multiplied with the variable.   **1c.** The degree is the same as the exponent on the variable.   **1d.** The degree is the sum of the exponents on the variables.   **2a.** A polynomial is the sum of two or more monomials.   **2b.** The degrees of the terms are in descending order.   **2c.** The degree of the polynomial is the same as the highest degree of any of the terms in the polynomial.   **3a.** monomial   **3b.** binomial   **3c.** trinomial   **3d.** polynomial   **4.** Like terms have the same variables, raised to the same powers.

5a. $9x^2 - x + 5 + 3x^2 + 1$ **5b.** $9x^2 + 3x^2 - x + 5 + 1$ **5c.** $(9+3)x^2 - x + (5+1)$ **5d.** $12x^2 - x + 6$

6a. $6z^3 + 2z^2 - 5 + 3z^3 - 9z^2 + z - 1$ **6b.** $6z^3 + 3z^3 + 2z^2 - 9z^2 + z - 5 - 1$ **6c.** $9z^3 - 7z^2 + z - 6$ **7a.** 117

7b. -91 **7c.** 26 **7d.** $2x^2 - 6$ **7e.** 26

Do the Math   **1.** $5, 4$ **2.** $-12, 2$ **3.** $-7, 0$ **4.** no   **5.** yes, 0, monomial   **6.** no   **7.** yes, 9, trinomial   **8.** $4y^4$

9. $-2t^3 - 3t^2 - 3t + 7$ **10.** $-10xy^2 - 4xy - y^2$ **11.** $\dfrac{5}{4}y^3 + \dfrac{7}{24}y - \dfrac{1}{6}$ **12.** $-w^3 + 4w^2 + 6w + 4$ **13.** -1 **14.** -11

15. 44 **16.** $6x$ **17.** 9 **18.** $3x^3 - 11x + 3$ **19.** $-7q^3 + 4q^2 - 5q + 1$ **20a.** \$29,192.94   **20b.** \$56,208.74

Section 4.2

Five-Minute Warm-Up   **1.** x^9 **2.** $21y^7$ **3.** $64z^3$ **4.** $16a^8$ **5.** $\dfrac{25}{4}x^2$ **6.** $\dfrac{2x^6}{9y}$ **7.** $21a - 14b$ **8.** $\dfrac{4x}{5} + \dfrac{3}{4}$

Guided Practice   **1.** The Distributive Property   **2.** $-2a^3b^2 - 14a^2b^3 + 6ab^4$ **3a.** The Distributive Property   **3b.** FOIL   **4.** $27x^2 + 39x - 10$ **5.** $12a^2 - 17ab - 7b^2$ **6a.** $(3x-1) \cdot 2x^2 + (3x-1) \cdot x + (3x-1) \cdot (-5)$

6b. $6x^3 - 2x^2 + 3x^2 - x - 15x + 5$ **6c.** $6x^3 + x^2 - 16x + 5$ **7.** $A^2 - B^2$ **8a.** $A^2 + 2AB + B^2$

8b. $A^2 - 2AB + B^2$ **9a.** -9 **9b.** -4 **9c.** 36 **9d.** $7x^3 - 2x^2 - 35x + 10$ **9e.** 36

Do the Math   **1.** $-27a^5b^7$ **2.** $9x^6y^4$ **3.** $-12m^3n^3 + 3m^2n^4 - 15mn^5$ **4.** $z^2 - 5z - 24$

5. $-14y^2 - 31y + 10$ **6.** $2y^2 + \dfrac{23}{6}y - 4$ **7.** $3m^2 + mn - 10n^2$ **8.** $21p^3 - 41p^2 + 31p - 6$

9. $x^3y + 2x^2y^2 + 4xy^3 - 2x^2 - 4xy - 8y^2$ **10.** $49p^2 - 42pq + 9q^2$ **11.** $m^4 - 4n^6$

12. $m^2 + 8m - 2mn + n^2 - 8n + 16$ **13.** $x^3 + 3x^2 - 7x + 15$ **14.** 48 **15.** $-2x^2 - 7x - 11$

16. $-4xh + h - 2h^2$ **17.** $-3x^3 + 18x^2 - 27x$ **18.** $12x^2 - 19x - 21$ **19.** $8y^3 + 27$ **20.** $9x^2 + 2x + \dfrac{1}{9}$

21. $z^3 - 9z^2 + 27z - 27$ **22.** $8a^2 - 18a - 2b^2 + 11b - 5$ **23.** $a^4 - a^3 - 11a^2 + 3a + 25$

Section 4.3

Five-Minute Warm-Up   **1.** r^5 **2.** $\dfrac{7x^4}{3}$ **3.** 25 **4.** $\dfrac{a^3}{2b^2}$ **5.** $\dfrac{81s^2}{4t^{10}}$ **6.** $\dfrac{2x^2y^2}{z}$ **7.** $\dfrac{4}{3}$

Guided Practice   **1a.** x^2 **1b.** $x^3 + x^2$ **1c.** See Graphing Answer Section

1d. $x^2 - 3x + 4 + \dfrac{2}{x+1}$ **2.** standard; 0   **3.** $x - c$; $x + c$ **4a.** $x^4 + 3x^3 - 9x^2 + 0 + 18$; $1, 3, -9, 0, 18$

4b. $3\overline{)1\ \ 3\ \ -9\ \ 0\ \ 18}$ **4c.** $\dfrac{3\overline{)1\ \ 3\ \ -9\ \ 0\ \ 18}}{1}$ **4d.** $\dfrac{3\overline{)1\ \ 3\ \ -9\ \ 0\ \ 18}\ \ \ 3}{1}$ **4e.** $\dfrac{3\overline{)1\ \ 3\ \ -9\ \ 0\ \ 18}\ \ \ 3}{1\ \ 6}$

4f. $\begin{array}{r} 3\overline{)1\ \ 3\ \ -9\ \ \ 0\ \ 18} \\ 3\ \ 18\ \ 27\ \ 81 \\ \hline 1\ \ 6\ \ \ 9\ \ 27\ \ 99 \end{array}$ **4g.** quotient: $x^3 + 6x^2 + 9x + 27$; remainder : 99   **4h.** $x^3 + 6x^2 + 9x + 27 + \dfrac{99}{x-3}$

5. $f(c)$ **6.** $f(c) = 0$

Sullivan/Struve, *Intermediate Algebra*, 3e
Copyright © 2014 Pearson Education, Inc.

Do the Math 1. $2z + 3$ **2.** $\dfrac{1}{2}z + \dfrac{2}{z}$ **3.** $y + \dfrac{9y}{2x} + \dfrac{8}{y}$ **4.** $x - 7$ **5.** $x - 6 + \dfrac{9}{4x + 7}$ **6.** $x^2 - x - 20$

7. $a^2 - 8a + 15$ **8.** $x - 5$ **9.** $k + 5 + \dfrac{-3k + 7}{2k^2 - 3}$ **10.** $x + 6 + \dfrac{7}{x - 4}$ **11.** $x^2 - 3x - 4 - \dfrac{5}{x + 3}$

12. $a^3 + 8a^2 - a - 8 - \dfrac{9}{a - 8}$ **13.** $\dfrac{3}{2}x - \dfrac{15}{4} + \dfrac{35}{4(2x + 1)}$ **14.** 1 **15.** -68 **16.** not a factor **17.** $(5x + 2)$ ft

18. $(x + 7)$ ft

Section 4.4

Five-Minute Warm-Up 1. $2 \cdot 2 \cdot 3 = 2^2 \cdot 3$ **2.** $2 \cdot 3 \cdot 3 = 2 \cdot 3^2$ **3.** $2^3 \cdot 3^2$ **4.** 5^3 **5.** $-24x + 16$

6. $7u^5 + 28u^4 - 7u^3$ **7.** 2, 3, 5, 7, 11, 13, 17, 19, 23, 29 **8a.** 12, 8 **8b.** 96 **9.** $3x^2 y$

Guided Practice 1a. $2m^2 n^2$ **1b.** $2m^2 n^2 \bullet 3m^2 + 2m^2 n^2 \bullet 9mn^2 - 2m^2 n^2 \bullet 11n^3$

1c. $2m^2 n^2\left(3m^3 + 9mn^2 - 11n^3\right)$ **2.** $7(a - 1)(a - 3)$ **3.** four **4.** true **5a.** $2x$ **5b.** $3y$

5c. $2x(x - 2) + 3y(x - 2)$ **5d.** $(x - 2)(2x + 3y)$

Do the Math 1. $8(z + 6)$ **2.** $-4(b - 8)$ **3.** $3a(4a + 15)$ **4.** $-6q\left(q^2 - 6q + 8\right)$

5. $-2b\left(9b^2 - 5b - 3\right)$ **6.** $(5z + 3)(6z + 5)$ **7.** $4ab^2\left(2a^3 + 3a^2 b - 9b^2\right)$ **8.** $(x - y)(8 + b)$

9. $(y + 3)\left(3y^2 - 5\right)$ **10.** $3(a - 5)(a - 3)$ **11.** $2y(y + 7)(y - 2)$ **12.** $6(x - 3)$ **13.** $(c - 1)\left(c^2 + 5\right)$

14. $2\pi r(r + 4)$ sq in **15a.** $0.08x$ **15b.** $0.80x - 0.15(0.80x)$ **15c.** $0.68x$ **15d.** \$442 **16.** (a) Receiving a

30% discount is the better deal. Explanations may vary.

Section 4.5

Five-Minute Warm-Up 1. 3, $-4, -7$ **2a.** -13 **2b.** 36 **3a.** -9 **3b.** -36 **4.** 7, -4 **5.** $-9, -6$

6. 12, 4 **7.** 21, -3 **8.** 4, 6 **9.** $-12, 3$

Guided Practice 1a. 19; 11; 9; -19; -11; -9 **1b.** 3 and 6 **1c.** $(y + 3)(y + 6)$ **2.** List all possible

combinations of factors of c and check to see if any sum to b **3a.** Grouping **3b.** Trial and Error **4a.** 3; 12

4b. 36 **4c.** -37; -20; -15; -13; -12 **4d.** -4 and -9 **4e.** $3x^2 - 4x - 9x + 12$

4f. $x(3x - 4) - 3(3x - 4) = (x - 3)(3x - 4)$ **5.** $2(4x + 3y)(6x - 5y)$ **6a.** p^2 **6b.** $12u^2 - u - 1$

6c. $(3u - 1)(4u + 1)$ **6d.** $\left(3p^2 - 1\right)\left(4p^2 + 1\right)$ **7a.** $-2\left(8x^2 + 2x - 3\right)$ **7b.** 8; 2; -3; -24; See Graphing Answer

Section **7c.** 6, -4 **7d.** $\dfrac{8}{6}; \dfrac{4}{3}$; 4; 3 **7e.** $\dfrac{8}{-4}; \dfrac{2}{-1}$; 2; -1 **7f.** $(4x + 3)(2x - 1)$ **7g.** $-2(4x + 3)(2x - 1)$

8. $-3n(2n - 3)(3n + 1)$

Do the Math 1. $(z + 7)(z - 4)$ **2.** $(q + 10)(q - 8)$ **3.** $-(p + 6)(p - 9)$ **4.** $(m + 2n)(m + 5n)$

5. $-4(s + 6)(s + 2)$ **6.** $(6x - 1)(x - 6)$ **7.** $(3r + 5)(4r - 3)$ **8.** $(4r + 3)(5r + 2)$ **9.** $(m + 3n)(3m - 2n)$

10. $4x(x - 9)(x - 4)$ **11.** $3xy(2x + 3)(9x - 8)$ **12.** $\left(y^2 + 3\right)\left(y^2 + 2\right)$ **13.** $(rs + 12)(rs - 4)$

14. $(3z + 11)(z + 7)$ **15.** $(4m + 9n)(6m + n)$ **16.** $\left(r^3 - 4\right)\left(r^3 - 2\right)$ **17.** $(p - 5q)(p - 9q)$ **18.** $(9a + 17)(a + 1)$

19. $(a + 3)(a - 2)$ **20.** $3(y + 2)(8y - 3)$ **21.** $-3mn(4m - 3n)(2m + 3n)$ **22.** $3xy(2x + 3)(9x - 8)$

Section 4.6

Five-Minute Warm-Up 1. 1, 4, 9, 16, 25, 36, 49, 64, 81 **2.** 1, 8, 27, 64 **3.** $\dfrac{25}{9}$ **4.** -64 **5.** $\dfrac{16}{9}$

6. $-\dfrac{125}{8}$ **7.** $64x^6 y^3$ **8.** $\dfrac{9}{4}a^2 b^6$ **9.** $9x^2 - 12x + 4$

Guided Practice 1a. $A^2 + 2AB + B^2$ **1b.** $A^2 - 2AB + B^2$ **2a.** p **2b.** 9 **2c.** $p^2 + 2(p \cdot 9) + 9^2$

2d. $(p + 9)^2$ **3.** 81 **4.** $A^2 - B^2$ **5a.** $(x - 8)(x + 8)$ **5b.** $\left(5m^3 - 6n^2\right)\left(5m^3 + 6n^2\right)$

Guided Practice **6a.** $(x^2 + 4x + 4) - 4y^2$ **6b.** $(x+2)^2 - (2y)^2$ **6c.** $(x+2+2y)(x+2-2y)$

7a. $A^3 + B^3$ **7b.** $A^3 - B^3$ **7c.** $A^3 + 3A^2B + 3AB^2 + B^3$ **8a.** p **8b.** 4 **8c.** $p^3 + 4^3$

8d. $(p+4)(p^2 - 4p + 16)$ **9a.** $3x(8x^3 + 125)$ **9b.** $3x((2x)^3 + 5^3)$ **9c.** $3x(2x+5)(4x^2 - 10x + 25)$

Do the Math **1.** $(3z-1)^2$ **2.** $(6b+7)^2$ **3.** $(2a+5b)^2$ **4.** $(b^2+4)^2$ **5.** $(9-a)(9+a)$

6. $(x^2 - 3y)(x^2 + 3y)$ **7.** $4z(3x - 4y)(3x + 4y)$ **8.** $(z+4)(z^2 - 4z + 16)$ **9.** $(6-n)(36 + 6n + n^2)$

10. $2(2m + 3n)(4m^2 - 6mn + 9n^2)$ **11.** $(5z+3)(7z^2 + 3z + 9)$ **12.** $(m^3 + n^4)(m^6 - m^3n^4 + n^8)$

13. $(3a - b)(3a + b)$ **14.** $(4x - 5)(16x^2 + 20x + 25)$ **15.** $3m(m - 3n)(m^2 + 3mn + 9n^2)$

16. $(p-3)^2(p+3)^2$ **17.** $(3mn - 5)^2$ **18.** $(p+4-q)(p+4+q)$ **19.** $-5(a+2)(a^2 - 2a + 4)$

20. $(6m + n - 9)(6m + n + 9)$ **21.** prime **22.** $(p-2+q)(p-2-q)$ **23.** $(x+0.3)^2$ **24.** $\left(\dfrac{a}{6} + \dfrac{b}{7}\right)\left(\dfrac{a}{6} - \dfrac{b}{7}\right)$

Section 4.7

Five-Minute Warm-Up **1.** $-8x^3y^7$ **2.** $6a^3b^2 - a^4b + \dfrac{5}{2}ab^2$ **3.** $20x^2 + 7xy - 6y^2$ **4.** $a^2 + 8ab + 16b^2$

5. $-9a(3a^2 - a + 2)$ **6.** $(4p + 3)(p - 2q)$

Guided Practice **1.** factor out the GCF **2a.** $-4x$ **2b.** $-4x(3x^2 + 5xy - 2y^2)$ **2c.** 3

2d. $-4x(3x - y)(x + 2y)$ **3a.** 2 **3b.** $(8a - 5b^2)(8a + 5b^2)$ **4a.** There is no GCF. **4b.**

$(-x^3 - 2x^2) + (5x + 10)$ **4c.** $-x^2(x + 2) + 5(x + 2)$ **4d.** $(x + 2)(-x^2 + 5)$ **5.** All factors are monomials or are prime.

Do the Math **1.** $3(x + 7)(x - 5)$ **2.** $-5(a + 4)(a - 4)$ **3.** $(4m - 7)(2m - 7)$

4. $2(3p^2 - q)(9p^4 + 3p^2q + q^2)$ **5.** $-4c(c^2 - 4c + 7)$ **6.** $(6t + 5)(3t - 4)$ **7.** $2(6p^2 + 25q^2)$

8. $(4w^2 + 1)(2w + 1)(2w - 1)$ **9.** prime **10.** $2pq(5p + 2)(2p - 1)$ **11.** $2p^2(3p + 2q)(9p^2 - 6pq + 4q^2)$

12. $(2z^2 + 5)(2z^2 - 5)$ **13.** $(2b^2 + 5)(2b^2 - 3)$ **14.** $3(x + 4)(3x + 2)$ **15.** $(a - 2b + 6)(a + 2b + 6)$

16. $(w^3 + 5)(w - 1)(w^2 + w + 1)$ **17.** $(q^2 + 1)(q^4 - q^2 + 1)$ **18.** $-2(y + 2)(y + 4)(y - 4)$

19. $-5z(1 + 4z^2)$ **20.** $2h(h^2 + 9)(3h - 2)(3h + 2)$ **21.** $f(x) = (x + 8)(x - 5)$ **22.** $H(p) = (5p - 2)(p + 6)$

23. $g(x) = -4(5x - 3)(5x + 3)$ **24.** $G(x) = (2x - 1)(x - 3)(x + 3)$

Section 4.8

Five-Minute Warm-Up **1.** $\left\{\dfrac{3}{2}\right\}$ **2.** $\left\{-\dfrac{1}{5}\right\}$ **3a.** 204 **3b.** $\dfrac{37}{4}$ **4.** $3; (3, -3)$ **5.** $-7; (-5, -7)$ **6.** $x = -\dfrac{3}{2}$

Guided Practice **1.** If the product of two numbers is zero, then one of the factors must be zero. That is, if $a \bullet b = 0$, then $a = 0$ or $b = 0$ or both a and b are zero. **2.** quadratic equation **3.** List the terms from the highest power to the lowest power. **4a.** $2x^2 - 3x - 2 = 0$ **4b.** $(2x + 1)(x - 2) = 0$ **4c.** $2x + 1 = 0$

4d. $x - 2 = 0$ **4e.** $x = -\dfrac{1}{2}$ **4f.** $x = 2$ **4g.** $\left\{-\dfrac{1}{2}, 2\right\}$ **5a.** Simplify the left side of the equation.

5b. Standard form **5c.** $\{10, -5\}$ **6a.** $p^3 + 2p^2 - 9p - 18 = 0$ **6b.** $(p - 3)(p + 3)(p + 2) = 0$

6c. $p - 3 = 0$ **6d.** $p + 3 = 0$ **6e.** $p + 2 = 0$ **6f.** $p = 3$ **6g.** $p = -3$ **6h.** $p = -2$ **6i.** $\{3, -3, -2\}$

7a. $3x^2 - 13x - 10 = 0$ **7b.** $-\dfrac{2}{3}$ and 5 **7c.** $\left(-\dfrac{2}{3}, 0\right)$ and $(5, 0)$ **8a.** $n + 2$ **8b.** $A = l \bullet w$ **8c.** $n; n + 2$

8d. $n(n + 2) = 255$ **8e.** $\{-17, 15\}$ **8f.** no; distance cannot be negative **8g.** The dimensions of the rectangle are: width of 15 cm by length of 17 cm.

Do the Math 1. $\left\{0, -\dfrac{4}{3}\right\}$ 2. $\{-11, 0, 9\}$ 3. $\{-3, 0\}$ 4. $\{-8, 5\}$ 5. $\{-6\}$ 6. $\left\{\dfrac{1}{4}, 6\right\}$ 7. $\left\{-\dfrac{1}{3}, -\dfrac{5}{2}\right\}$

8. $\left\{-4, \dfrac{5}{2}\right\}$ 9. $\left\{-5, \dfrac{3}{2}\right\}$ 10. $\{-9, 5\}$ 11. $\{-5, -4, 4\}$ 12. $\left\{-\dfrac{3}{2}, 0, \dfrac{3}{4}\right\}$ 13. $\{-8, 4\}$ 14. $\{0, 6\}$

15. $x = 0$ or $x = -5$ 16. $x = 2$ or $x = -7$ 17. $\left(-\dfrac{5}{4}, 0\right)$ and $\left(\dfrac{7}{2}, 0\right)$ 18. height: 12 m; base: 8 m 19. 3 ft

20. width: 13 in.; length: 21 in.

Chapter 5 Answers
Section 5.1

Five-Minute Warm-Up 1. $(x - 5)(x + 2)$ 2. $(2z - 1)(3z + 2)$ 3. $\{-4, -5\}$ 4. $\dfrac{2}{9}$ 5. (a); (b) 6. $\dfrac{2}{5}$ 7. $\dfrac{9}{20}$

Guided Practice 1. the denominator to equal zero. 2a. $\{x \mid x \neq -2\}$ 2b. $\{a \mid a \neq 3, a \neq 6\}$

3a. $\dfrac{(x - 10)(x + 2)}{(x + 2)(2x - 1)}$ 3b. $\dfrac{x - 10}{2x - 1}$ 4a. $\dfrac{x(x - 4)}{(x - 2)(x + 2)} \cdot \dfrac{(x - 3)(x + 2)}{(x - 4)}$ 4b. $\dfrac{x(x - 4)(x - 3)(x + 2)}{(x - 2)(x + 2)(x - 4)}$

4c. $\dfrac{x\cancel{(x-4)}(x - 3)\cancel{(x+2)}}{(x - 2)\cancel{(x+2)}\cancel{(x-4)}}$ 4d. $\dfrac{x(x - 3)}{(x - 2)}$ 5. $\dfrac{x}{y} \div \dfrac{p}{q} = \dfrac{x}{y} \cdot \dfrac{q}{p}$ 6a. $\dfrac{27xy^2}{60x^2y^4} \cdot \dfrac{24y^5}{36x}$

6b. $\dfrac{27 \cdot 24 \cdot x \cdot y^2 \cdot y^5}{60 \cdot 36 \cdot x^2 \cdot x \cdot y^4}$ 6c. $\dfrac{3y^3}{10x^2}$ 7. $q(x) = 0$ 8. $\{x \mid x \neq -8, x \neq 5\}$

Do the Math 1. $\{x \mid x \neq 7\}$ 2. $\{x \mid x \text{ is any real number}\}$ 3. $\{x \mid x \neq -4\}$ 4. $\dfrac{x}{x + 3}$ 5. $\dfrac{w + 7}{w + 8}$

6. $\dfrac{x - 3y}{x - 2y}$ 7. $\dfrac{v^2 - 5}{v + 3}$ 8. $\dfrac{x(x + 4)}{4}$ 9. $\dfrac{x + 5}{x + 6}$ 10. $-4y - 3$ 11. $-\dfrac{1}{3}$ 12. $\dfrac{(x - 3)^2}{x(2x - 5)}$ 13. $\dfrac{5(m - 4)}{m(m + 4)}$

14. $\dfrac{1}{15}$ 15. $12m^2n^2$ 16. $\dfrac{y - 2}{3y + 1}$ 17. $\dfrac{1}{x - y}$ 18. $\left\{x \mid x \neq -5, x \neq \dfrac{1}{4}\right\}$ 19. $\{x \mid x \text{ is any real number}\}$

20a. $\dfrac{(x + 1)(x + 4)}{x - 2}$ 20b. $\dfrac{(x + 1)(x + 9)}{(2x - 5)(x + 5)}$

Section 5.2

Five-Minute Warm-Up 1. 21 2. 600 3. $\dfrac{7}{24} = \dfrac{175}{600}; \dfrac{14}{75} = \dfrac{112}{600}$ 4. 6 5. $\dfrac{89}{90}$ 6. $-\dfrac{3}{14}$ 7. $\dfrac{25}{48}$

Guided Practice 1a. $\dfrac{x^2 - 3 + (x^2 + x)}{2x + 3}$ 1b. $\dfrac{2x^2 + x - 3}{2x + 3}$ 1c. $\dfrac{(2x + 3)(x - 1)}{(2x + 3)}$ 1d. $x - 1$

2a. $\dfrac{x^2 - 11 + (-1)(-3x - 1)}{x^2 - 25}$ 2b. $\dfrac{x^2 - 11 + 3x + 1}{x^2 - 25}$ 2c. $\dfrac{x^2 + 3x - 10}{x^2 - 25}$ 2d. $\dfrac{(x + 5)(x - 2)}{(x + 5)(x - 5)}$ 2e. $\dfrac{x - 2}{x - 5}$

3. **Step 1:** Factor each denominator completely. When factoring, write the factored form using powers.
Step 2: The LCD is the product of each prime factor the greatest number of times it appears in any factorization. 4. **Step 1:** Find the LCD. **Step 2:** Rewrite each rational expression as an equivalent rational expression with the common denominator. You will need to multiply out the numerator, but leave the denominator in factored form. **Step 3:** Add or subtract the rational expressions found in Step 2.

Step 4: Simplify the result. 5a. $(x - 2)(x + 2)$ 5b. $\dfrac{3}{x + 2} = \dfrac{3x - 6}{(x + 2)(x - 2)}; \dfrac{8 - 2x}{(x + 2)(x - 2)}$ 5c. $\dfrac{1}{x - 2}$.

6a. $(a + 10)(a + 2)(a - 2)$ **continued next page**

Guided Practice 6b. $\dfrac{a}{(a+10)(a+2)}=\dfrac{a^2-2a}{(a+10)(a+2)(a-2)}$; $\dfrac{1}{(a+10)(a-2)}=\dfrac{a+2}{(a+10)(a-2)(a+2)}$

6c. $\dfrac{a^2-3a-2}{(a+10)(a+2)(a-2)}$ **7a.** $x^2(x-1)^2$ **7b** $\dfrac{x}{(x-1)^2}=\dfrac{x^3}{x^2(x-1)^2}$; $\dfrac{2}{x}=\dfrac{2x^3-4x^2+2x}{x^2(x-1)^2}$;

$\dfrac{x+1}{(-1)x^2(x-1)}=\dfrac{-x^2+1}{x^2(x-1)^2}$ **7c.** $\dfrac{-x^3+5x^2-2x-1}{x^2(x-1)^2}$

Do the Math 1. $\dfrac{5x+2}{x-3}$ **2.** $\dfrac{9x-2}{6x-5}$ **3.** $\dfrac{3x-2}{x-6}$ **4.** $2x+5$ **5.** $24a^3b^2$ **6.** $(m+6)(m-3)(m-4)$

7. $x^2(x-3)(x+3)$ **8.** $\dfrac{2x+15}{9x^2}$ **9.** $\dfrac{4(x+2)}{(x-3)(x+1)}$ **10.** $\dfrac{z-7}{z-4}$ **11.** $\dfrac{2x^2-5x-1}{(x-1)(x+1)(x+3)}$

12. $\dfrac{2m^2-4mn+4n^2}{(m+2n)^2(m-3n)}$ **13.** $-\dfrac{1}{(x+1)(x+2)}$ **14.** 0 **15.** $\dfrac{15}{m(m-3)}$ **16.** $\dfrac{2(3x+4)}{x(x+2)^2}$ **17.** $\dfrac{4(x-2)}{x-4}$

18a. $\dfrac{8x+1}{(x+2)(x-1)}$ **18b.** $\{x\mid x\neq -2,\ x\neq 1\}$ **19a.** $S(r)=\dfrac{2\pi r^3+400}{r}$ **19b.** $S(4)\approx 200.53$ sq cm

Section 5.3

Five-Minute Warm-Up   1. $(4x+3)(2x-1)$ **2.** $\dfrac{4p^3}{5}$ or $\dfrac{4}{5}p^3$ **3.** $\dfrac{9y}{20x^3}$ **4.** $16r^{12}$ **5.** $\dfrac{1}{y^8z^5}$ **6.** $\dfrac{4}{9}$ **7.** $-\dfrac{7}{8}$

Guided Practice   1. When sums and/or differences of rational expressions occur in the numerator or denominator of a quotient, the quotient is called a complex rational expression. Rational expressions with more than one fraction bar are complex and need to be simplified. **2a.** $x-1$ **2b.** $\dfrac{x-1+1}{x-1}=\dfrac{x}{x-1}$

2c. $\dfrac{\dfrac{x}{x+1}}{\dfrac{x}{x-1}}$ **2d.** $\dfrac{x}{x+1}\cdot\dfrac{x-1}{x}$ **2e.** $\dfrac{x-1}{x+1}$ **3a.** Multiplicative Inverse (a number multiplied by its reciprocal is one) or alternately, any quotient with the same numerator and denominator is a representation of one whole.

3b. Multiplicative Identity   **3c.** Distributive Property   **4a.** $(x-4)(x+4)$

4b. $\left[\dfrac{\dfrac{x^2}{(x-4)(x+4)}-\dfrac{x}{(x+4)}}{\dfrac{x}{(x-4)(x+4)}-\dfrac{1}{(x-4)}}\right]\left[\dfrac{(x-4)(x+4)}{(x-4)(x+4)}\right]$ **4c.** $\dfrac{\dfrac{x^2(x-4)(x+4)}{(x-4)(x+4)}-\dfrac{x(x-4)(x+4)}{(x+4)}}{\dfrac{x(x-4)(x+4)}{(x-4)(x+4)}-\dfrac{1\cdot(x-4)(x+4)}{(x-4)}}$ **4d.** $\dfrac{x^2-x(x-4)}{x-(x+4)}$

4e. $\dfrac{x^2-x^2+4x}{x-x-4}=\dfrac{4x}{-4}=-x$ **5a.** $\dfrac{\dfrac{1}{x}+\dfrac{1}{3}}{\dfrac{1}{x^2}-\dfrac{1}{9}}$ **5b.** $\dfrac{3x}{3-x}$ or $-\dfrac{3x}{x-3}$

Do the Math 1. $\dfrac{x^2+1}{(x+1)(x-1)}$ **2.** $\dfrac{7x+9w}{9x-7w}$ **3.** $\dfrac{a}{(a-3)(a+1)}$ **4.** $\dfrac{1}{(x-2)(x-1)}$ **5.** $\dfrac{-2(x-2)}{(x+1)(x-1)}$

6. $\dfrac{1}{z+4}$ **7.** n^2+nm+m^2 **8.** $\dfrac{x+4}{x}$ **9.** $-x$ **10.** $\dfrac{x-3}{x+3}$ **11.** $\dfrac{6}{x+5}$ **12.** $\dfrac{2}{x-y}$ **13.** $-\dfrac{xy}{(x-y)^2}$

14. $\dfrac{b^2-2ab+4a^2}{ab(b-2a)}$ **15a.** $\dfrac{R_1R_2R_3}{R_2R_3+R_1R_3+R_1R_2}$ **15b.** $\dfrac{60}{31}\approx 1.94$ ohms

Sullivan/Struve, *Intermediate Algebra*, 3e
Copyright © 2014 Pearson Education, Inc.

Section 5.4

Five-Minute Warm-Up **1.** $\{4\}$ **2.** $\{-4, 6\}$ **3.** $(3z - 2)(z - 4)$ **4.** (b) only **5.** $\{15, -3\}$ **6.** $(-2, 12)$

Guided Practice **1.** the denominator equal to zero **2a.** $x = 7, x = -7$ **2b.** $\{x \mid x \neq 7, x \neq -7\}$

2c. $(x - 7)(x + 7)$ **2d.** $(x - 7)(x + 7)\dfrac{x + 5}{x - 7} = \dfrac{x - 3}{x + 7}(x - 7)(x + 7)$ **2e.** $(x + 7)(x + 5) = (x - 3)(x - 7)$

2f. $x^2 + 12x + 35 = x^2 - 10x + 21$ **2g.** $x = -\dfrac{7}{11}$ **2h.** $\left\{-\dfrac{7}{11}\right\}$ **3a.** $\{x \mid x \neq 0\}$ **3b.** $12x^2$

3c. $x^3 + 6x^2 = 4x + 24$ **3d.** $\{2, -2, -6\}$ **4a.** $\{x \mid x \neq 1, x \neq -1\}$ **4b.** $(x - 1)(x + 1)$ **4c.** $x = -1$ **4d.** $\varnothing$

5a. $\{x \mid x \neq 4, x \neq 2\}$ **5b.** $(x - 4)(x - 2)$ **5c.** $x = 5$ or $x = 4$ **5d.** $\{5\}$ **6a.** $x + \dfrac{7}{x} = 8$ **6b.** $\{x \mid x \neq 0\}$

6c. x **6d.** $x = 7$ or $x = 1$ **6e.** $\{7, 1\}$ **6f.** $(7, 8)$ and $(1, 8)$

Do the Math **1.** $\{6\}$ **2.** $\varnothing$ **3.** $\{2, 4\}$ **4.** $\left\{-\dfrac{1}{2}, \dfrac{3}{4}\right\}$ **5.** $\left\{-\dfrac{3}{2}, 4\right\}$ **6.** $\{0, 8\}$ **7.** $\{2\}$ **8.** $\varnothing$ **9.** $\{5\}$

10. $\{18\}$ **11.** $\{14\}$ **12.** $\left\{\dfrac{1}{2}, -\dfrac{1}{2}\right\}$ **13.** $\left\{-4, -\dfrac{3}{2}\right\}$ **14.** $x = -1, x = -4, (-1, -10), (-4, -10)$

15. $x = -13, \left(-13, \dfrac{17}{33}\right)$ **16.** $\dfrac{5z - 2}{z(z - 1)}$ **17.** $\left\{\dfrac{3}{7}\right\}$ **18.** $\left\{-\dfrac{1}{2}, 3\right\}$ **19.** $\dfrac{-2x^2 + 6x + 5}{(x - 2)(x + 1)}$

Section 5.5

Five-Minute Warm-Up **1.** $[-3, 2)$ **2.** $\{x \mid x < -1\}$ See Graphing Answer Section **3.** $\left[-\dfrac{5}{3}, \infty\right)$ **4.** yes

Guided Practice **1a.** $0; 0$ **1b.** $0; 0$ **2.** positive; negative **3.** positive; negative; negative; positive
4. denominator to equal zero **5a.** $x = -2$ **5b.** $x = 4$ **5c.** -3 (answers may vary) **5d.** negative
5e. negative **5f.** positive **5g.** 0 (answers may vary) **5h.** positive **5i.** negative **5j.** negative
5k. 5 (answers may vary) **5l.** positive **5m.** positive **5n.** positive **5o.** negative **5p.** $x = 4$

5q. See Graphing Answer Section **6.** $\dfrac{-2x + 12}{(x - 1)(x + 1)} > 0$ **7.** $\dfrac{x - 11}{x + 2} < 0$ **8.** See Graphing Answer Section

9. See Graphing Answer Section

Do the Math **1.** $(-\infty, -8) \cup (-2, \infty)$ **2.** $(-\infty, -12] \cup (2, \infty)$ **3.** $(-5, 10]$ **4.** $(-\infty, -4) \cup \left(\dfrac{2}{5}, 5\right)$

5. $\left(-1, \dfrac{2}{3}\right] \cup [6, \infty)$ **6.** $(4, \infty)$ **7.** $(-2, 11]$ **8.** $(-\infty, -6) \cup (-5, \infty)$ **9.** $(-\infty, -3) \cup \left[-\dfrac{3}{2}, 0\right)$

10. $\left(-\infty, -\dfrac{1}{2}\right) \cup (4, 13]$ **11.** $(-\infty, -3] \cup (8, \infty)$ **12.** $\left(-\dfrac{2}{3}, 4\right)$ **13.** $(-4, 0)$ and $\left(\dfrac{5}{3}, 0\right)$ **14.** $(-3, 0)$

15. 125 or more bicycles

Section 5.6

Five-Minute Warm-Up **1.** $x = -\dfrac{3}{4}y - 6$ **2.** $y = -\dfrac{9}{2}x + 18$ **3.** 65 mph

Guided Practice **1.** $y = \dfrac{xz}{2z - 6x}$ **2.** $\dfrac{3}{x}$ Answers may vary **3.** $\dfrac{3}{x} = \dfrac{x + 2}{9}$ Answers may vary **4.** $\left\{\dfrac{13}{3}\right\}$

5. The figures are the same shape, but they are a different size. The corresponding angles are equal and the

corresponding sides are proportional. **6a.** $\dfrac{6}{3.2} = \dfrac{x}{8}$ **6b.** 15 ft **7a.** $\dfrac{1}{12}$ **7b.** $\dfrac{1}{x}$ **7c.** $\dfrac{1}{t + 2}$ **8a.** $\dfrac{1}{3}$ **8b.** $\dfrac{1}{5}$

8c. $\dfrac{1}{t}$ **8d.** $\dfrac{1}{3} + \dfrac{1}{5} = \dfrac{1}{t}$ **8e.** It takes 1.875 hours to clean the building when Josh and Ken work together.

Guided Practice **9a.** $180 + w$; where w is the speed of the wind **9b.** $180 - w$ **9c.** $\dfrac{1000}{180 + w}$

9d. $\dfrac{600}{180 - w}$ **9e.** $\dfrac{1000}{180 + w} = \dfrac{600}{180 - w}$ **9f.** The speed of the wind is 45 mph.

Do the Math **1.** $V_2 = \dfrac{V_1 P_1}{P_2}$ **2.** $r = \dfrac{A}{P} - 1$ **3.** $x_1 = \dfrac{y_1 - y + mx}{m}$ **4.** $m_1 = \dfrac{m_2 v_2}{2v_1 - v_2}$

5. about 21.02 million flight hr **6.** approximately 5.83 hr **7.** 15 hours **8.** 3 minutes; $\dfrac{1}{2}$ mile **9.** 6 hours

10. 135 miles per hour **11.** 10 miles per hour

Section 5.7

Five-Minute Warm-Up **1a.** $\{-5\}$ **1b.** $\left\{-\dfrac{4}{9}\right\}$ **2.** $\{-48\}$ **3 – 4.** See Graphing Answer Section

Guided Practice **1.** $y = kx$ **2.** constant of proportionality **3.** linear; zero **4.** $y = -\dfrac{25}{3}$ **5.** $y = \dfrac{k}{x}$

6. $y = \dfrac{8}{3}$ **7.** $r = 42$ **8a.** $V = \dfrac{k \cdot T}{P}$ **8b.** $T = 300$; $P = 15$; $V = 100$ **8c.** $k = 5$ **8d.** $V = \dfrac{5 \cdot T}{P}$ **8e.** 22.5 Atm

Do the Math **1a.** $k = 5$ **1b.** $y = 5x$ **1c.** $y = 25$ **2a.** $k = \dfrac{1}{5}$ **2b.** $y = \dfrac{1}{5}x$ **2c.** $y = 7$ **3.** $B = 80$

4. $r = 64$ **5.** \$1225.62 **6.** \$80.50 **7.** 96 feet per second **8a.** $k = 80$ **8b.** $y = \dfrac{80}{x}$ **8c.** $y = \dfrac{16}{7}$ **9a.** $k = \dfrac{1}{3}$

9b. $y = \dfrac{1}{3}xz$ **9c.** $y = 40$ **10a.** $k = \dfrac{21}{10}$ **10b.** $Q = \dfrac{21x}{10y}$ **10c.** $Q = \dfrac{28}{5}$ **11.** 24 amps **12.** ≈ 0.033 foot-candles

Chapter 6 Answers
Section 6.1

Five-Minute Warm-Up **1.** 0.5 **2.** $\dfrac{4}{9}$ **3.** $|4x + 3|$ **4.** $|x - y|$ **5.** $\dfrac{1}{49}$ **6.** x^2 **7.** $\dfrac{9x^2}{y^4}$ **8.** $8b^{12}$

Guided Practice **1.** index; radicand; cube root **2.** ≥ 0; any real number **3.** principal root **4a.** -3

4b. not a real number **4c.** $\dfrac{2}{3}$ **5.** ≈ 2.83 **6.** $|2x - 1|$ **7a.** $\sqrt{144} = 12$ **7b.** $\sqrt{-64}$ not a real number

7c. $2\sqrt[3]{x}$ **8.** $\sqrt[n]{a^m}$; $\left(\sqrt[n]{a}\right)^m$ **9a.** 216 **9b.** -1024 **9c.** 16 **10.** ≈ 8.55 **11a.** $(2x)^{\frac{3}{4}}$ **11b.** $\left(2x^2 y\right)^{\frac{2}{3}}$

12a. $\dfrac{1}{64^{\frac{1}{2}}} = \dfrac{1}{8}$ **12b.** $2 \cdot 16^{\frac{1}{2}} = 2\sqrt{16} = 8$ **12c.** $\dfrac{1}{(4x)^{\frac{5}{2}}} = \dfrac{1}{32x^{\frac{5}{2}}}$

Do the Math **1.** 6 **2.** -4 **3.** -4 **4.** $\dfrac{2}{5}$ **5.** 6 **6.** n **7.** $|2x - 3|$ **8.** 4 **9.** -5 **10.** -3

11. not a real number **12.** $-100{,}000$ **13.** 8 **14.** $\dfrac{1}{11}$ **15.** 343 **16.** $\dfrac{1}{81}$ **17.** $x^{\frac{3}{4}}$ **18.** $(3x)^{\frac{2}{5}}$

19. $(3pq)^{\frac{7}{4}}$ **20.** 4.40 **21.** 3.16 **22.** 1000 **23.** not a real number **24.** $\dfrac{1}{5}$ **25.** 2 **26.** $-\dfrac{1}{5}$ **27.** $(-9)^{\frac{1}{2}} = \sqrt{-9}$

but there is no real number whose square is -9. However, $-9^{\frac{1}{2}} = -1 \cdot 9^{\frac{1}{2}} = -\sqrt{9} = -3$.

Section 6.2

Five-Minute Warm-Up **1.** $54z^7$ **2.** $\dfrac{3u^2}{2}$ or $\dfrac{3}{2}u^2$ **3.** $\dfrac{1}{625}$ **4.** $\dfrac{27}{64}$ **5.** $\dfrac{3x^6}{y}$ **6.** $\dfrac{1}{25p^6}$ **7.** $\dfrac{64b^{10}}{a^2}$

8. $\dfrac{1}{27xy^7}$ **9.** $\dfrac{9}{7}$

Guided Practice 1. All exponents are positive; each base only occurs once; there are no parentheses in the expression; there are no powers written to powers. **2a.** 36 **2b.** $-2a^{\frac{1}{2}}$ **3.** $\dfrac{x^{\frac{11}{8}}}{y^{\frac{15}{8}}}$ **4a.** 9 **4b.** $2x^2|y|\sqrt[4]{8}$

4c. $\sqrt[4]{x}$ **4d.** $\sqrt[6]{p}$ **5.** $2x^{\frac{1}{3}}\left(3x^{\frac{1}{3}}+5x-2\right)$

Do the Math 1. 9 **2.** 10 **3.** $\dfrac{1}{y^{\frac{7}{10}}}$ **4.** $\dfrac{2}{9}$ **5.** $a^{\frac{1}{2}}b^{\frac{3}{5}}$ **6.** $\dfrac{b^2}{a^{\frac{2}{3}}}$ **7.** $\dfrac{5p^{\frac{1}{5}}}{q^{\frac{1}{2}}}$ **8.** $\dfrac{8m^{\frac{5}{4}}}{n^{\frac{1}{6}}}$ **9.** $\dfrac{x^{\frac{1}{3}}}{y^{\frac{1}{9}}}$

10. $x^2+4x^{\frac{1}{3}}$ **11.** $\dfrac{6}{a^{\frac{1}{2}}}-3a^{\frac{1}{2}}$ **12.** $8p^2-32$ **13.** x^2 **14.** 25 **15.** $5x^2y^3$ **16.** $\sqrt[12]{p^{13}}$ **17.** $\sqrt[6]{5^7}$

18. $3(x-5)^{\frac{1}{2}}(5x-9)$ **19.** $\dfrac{2(6x+1)}{x^{\frac{2}{3}}}$ **20.** $\dfrac{6x+13}{(x+3)^{\frac{1}{2}}}$ **21.** $3(x^2-1)^{\frac{1}{3}}(3x-1)(x+3)$ **22.** 3 **23.** 125 **24.** 32 **25.** 0

Section 6.3
Five-Minute Warm-Up 1. 1, 4, 9, 16, 25, 36, 49, 64, 81, 100, 121, 144 **2.** 1, 8, 27, 64, 125 **3.** 1, 16, 81
4. not a real number **5.** 10 **6.** $|x|$ **7.** $|4x+1|$ **8.** $|2p-1|$ **9.** 13

Guided Practice 1a. $\sqrt{35}$ **1b.** $\sqrt{x^2-25}$ **1c.** $\sqrt[3]{14x^2}$ **2.** the radicand does not contain factors that are perfect powers of the index. For square roots, there can be no factor with exponent two or higher in the radicand. For cube roots, there can be no factor with exponent three or higher in the radicand. **3a.** 25

3b. $\sqrt{25}\cdot\sqrt{3}$ **3c.** $5\sqrt{3}$ **4a.** $16\cdot3$ **4b.** $\sqrt{16\cdot3}=\sqrt{16}\cdot\sqrt{3}$ **4c.** $\dfrac{-6+4\cdot\sqrt{3}}{2}$ **4d.** $\dfrac{2(-3+2\sqrt{3})}{2}$ **4e.** $-3+2\sqrt{3}$
5. Rewrite the variable expression as the product of two variable expressions, where one of the factors has an exponent that is a multiple of the index. **6.** $\dfrac{2x\sqrt{3}}{11}$ **7.** $\dfrac{6}{a^2}$ **8a.** Rewrite $\sqrt{6}\bullet\sqrt[4]{48}$ as $6^{\frac{1}{2}}\bullet48^{\frac{1}{4}}$. **8b.** 4

8c. $\left(6^2\right)^{\frac{1}{4}}\cdot(48)^{\frac{1}{4}}$ **8d.** $\left(6^2\cdot48\right)^{\frac{1}{4}}$ **8e.** $1728^{\frac{1}{4}}$ **8f.** $\sqrt[4]{2^6\cdot3^3}=2\sqrt[4]{108}$

Do the Math 1. $\sqrt[4]{42a^2b^2}$ **2.** $\sqrt{p^2-25}$; $|p|\ge5$ **3.** $\sqrt[3]{-3x}$ **4.** $3\sqrt[4]{2}$ **5.** $2|a|\sqrt{5}$ **6.** $-4p$ **7.** $12\sqrt{3b}$

8. $s^4\sqrt{s}$ **9.** $x^2\sqrt[5]{x^2}$ **10.** $-3q^4\sqrt[3]{2}$ **11.** $5|x^3|\sqrt{3y}$ **12.** -1 **13.** $\dfrac{-3+2\sqrt{3}}{4}$ **14.** 6 **15.** $6x\sqrt{5}$ **16.** $3a\sqrt[3]{2}$

17. $6m\sqrt[3]{2mn^2}$ **18.** $\dfrac{\sqrt{5}}{6}$ **19.** $\dfrac{x\sqrt[4]{5}}{2}$ **20.** $-\dfrac{3x^3}{4y^4}$ **21.** 2 **22.** $3y^2\sqrt{2}$ **23.** $-4x^3$ **24.** $\sqrt[6]{392}$ **25.** $\sqrt[8]{45}$

Section 6.4
Five-Minute Warm-Up 1. $3z^3-5z^2-4z$ **2.** $-3a^2b^2-a^2+6ab-b^2$ **3.** $-8x^4-12x^3y+20x^2y^2$
4. $\dfrac{2}{9}x^5-\dfrac{4}{7}x^4+x^2$ **5.** $18c^2-23c-6$ **6.** a^2b^2-4 **7.** $49n^2-42n+9$

Guided Practice 1. Like radicals have the same index and the same radicand. **2a.** $9\sqrt{x}$ **2b.** $8\sqrt[3]{3p}$
3a. $-10\sqrt{2}$ **3b.** $-n^4\sqrt{6}+4n^4\sqrt{6n}$ **4.** $18+4\sqrt{6}-27\sqrt{2}-12\sqrt{3}$ **5a.** $(A+B)(A-B)=A^2+B^2$
5b. $49-3=46$ **6.** 72 square units
Do the Math 1. $14\sqrt{3}$ **2.** $7\sqrt[4]{z}$ **3.** $11\sqrt[3]{5}-11\sqrt{5}$ **4.** $8\sqrt{3}$ **5.** $2\sqrt[4]{3}$ **6.** $3\sqrt{3z}$ **7.** $13z\sqrt{7z}$ **8.** $2y\sqrt{3}$
9. $3(2x-1)\sqrt[3]{5}$ **10.** $5\sqrt{5}+3\sqrt{15}$ **11.** $\sqrt[3]{12}+2\sqrt[3]{9}$ **12.** $15+3\sqrt{5}+5\sqrt{6}+\sqrt{30}$ **13.** $-141-22\sqrt{10}$
14. $-2\sqrt{3}$ **15.** $7-4\sqrt{3}$ **16.** 2 **17.** 18 **18.** $\sqrt[3]{y^2}-3\sqrt[3]{y}-18$ **19.** $-2\sqrt{3}$ **20.** $z+2\sqrt{5z}+5$ **21.** $8-4\sqrt{15}$

Do the Math **22.** $19+2x+8\sqrt{2x+3}$ **23.** $3a$ **24.** $9-4\sqrt{14}$ **25.** $\dfrac{2\sqrt{5}}{25}$

Section 6.5

Five-Minute Warm-Up **1.** 3 **2.** $11a$ **3.** $\dfrac{3\sqrt{2}}{2}$ **4.** 1 **5.** $24-\sqrt{15}$

Guided Practice **1.** The process of rationalizing the denominator of a quotient requires rewriting the quotient, using properties of rational expressions, so that the denominator of the equivalent expression does not contain any radicals. **2.** a perfect square **3a.** $\dfrac{\sqrt{3}}{\sqrt{3}}$ **3b.** $\dfrac{\sqrt{5}}{\sqrt{5}}$ **3c.** $\dfrac{\sqrt{2a}}{\sqrt{2a}}$ **4a.** $\dfrac{\sqrt[3]{25}}{\sqrt[3]{25}}$ **4b.** $\dfrac{\sqrt[3]{18}}{\sqrt[3]{18}}$ **4c.** $\dfrac{\sqrt[4]{3x^2y^3}}{\sqrt[4]{3x^2y^3}}$

5. conjugate **6a.** $3-\sqrt{5};\ 4$ **6b.** $2\sqrt{3}+5\sqrt{2};\ -38$ **7.** $\dfrac{\sqrt{2}+3}{\sqrt{2}+3}$ **8.** $-4\left(\sqrt{2}+\sqrt{5}\right)$

Do the Math **1.** $\dfrac{2\sqrt{3}}{3}$ **2.** $-\dfrac{\sqrt{3}}{2}$ **3.** $\dfrac{\sqrt{5}}{2}$ **4.** $\dfrac{\sqrt{33}}{11}$ **5.** $\dfrac{\sqrt{5z}}{z}$ **6.** $\dfrac{4\sqrt{2a}}{a^3}$ **7.** $-\dfrac{\sqrt[3]{4p^2}}{p}$ **8.** $-\dfrac{\sqrt[3]{15}}{6}$

9. $\dfrac{4\sqrt[3]{6z}}{3z}$ **10.** $\dfrac{2\sqrt[4]{9b^2}}{b}=\dfrac{2\sqrt{3b}}{b}$ **11.** $2\left(\sqrt{7}+2\right)$ **12.** $10\left(\sqrt{10}-3\right)$ **13.** $\dfrac{\sqrt{5}+\sqrt{2}}{3}$ **14.** $2+\sqrt{3}$ **15.** $\dfrac{4\sqrt{5}}{5}$

16. $\dfrac{11\sqrt{5}}{10}$ **17.** $\dfrac{\sqrt{10}-5\sqrt{5}}{5}$ **18.** $\sqrt{3}$ **19.** $\dfrac{1}{3}$ **20.** $\dfrac{3\sqrt{5}}{5}$ **21.** $\dfrac{10\sqrt{2}-27}{23}$ **22.** $\dfrac{4-\sqrt{6}}{2}$ **23.** 5

24. $\dfrac{1}{2\left(2-\sqrt{3}\right)}$ **25.** $\dfrac{a-b}{\sqrt{2a}+\sqrt{2b}}$

Section 6.6

Five-Minute Warm-Up **1.** 12 **2.** n **3.** $\{x\,|\,x\le-3\}$ **4.** -40 **5.** See Graphing Answer Section

Guided Practice **1a.** $4\sqrt{5}$ **1b.** $\sqrt[3]{-9}$ **1c.** $\sqrt{2}$ **2.** ≥ 0 **3.** any real number **4a.** $\left\{x\,\middle|\,x\ge\dfrac{3}{2}\right\}$ or $\left[\dfrac{3}{2},\infty\right)$

4b. $\{x\,|\,x$ is any real number$\}$ or $(-\infty,\infty)$ **4c.** $\{t\,|\,t\le 2\}$ or $(-\infty,2]$ **5a.** $[-3,\infty)$ **5b.** See Graphing Answer Section **5c.** $[0,\infty)$ **6a.** $(-\infty,\infty)$ **6b.** See Graphing Answer Section **6c.** $(-\infty,\infty)$

Do the Math **1a.** 4 **1b.** $2\sqrt{3}$ **1c.** 2 **2a.** 1 **2b.** -3 **2c.** $2\sqrt[3]{2}$ **3a.** $\dfrac{1}{3}$ **3b.** $\dfrac{\sqrt{3}}{3}$ **3c.** $\dfrac{\sqrt{2}}{2}$ **4.** $[-4,\infty)$

5. $\left(-\infty,\dfrac{5}{2}\right]$ **6.** $(-\infty,\infty)$ **7.** $\left[\dfrac{2}{3},\infty\right)$ **8.** $(3,\infty)$ **9a.** $[1,\infty)$ **9b.** $[0,\infty)$ **10a.** $(-\infty,4]$ **10b.** $[0,\infty)$

11a. $[0,\infty)$ **11b.** $[1,\infty)$ **12a.** $(-\infty,\infty)$ **12b.** $(-\infty,\infty)$ **9 – 12(c).** See Graphing Answer Section

Section 6.7

Five-Minute Warm-Up **1.** $\{-3\}$ **2.** $\{1,2\}$ **3.** $2x$ **4.** $2x+10\sqrt{2x}+25$ **5.** $3x+8$ **6.** $4\sqrt{3}$

Guided Practice **1a.** $\sqrt{4x+1}=5$ **1b.** $\left(\sqrt{4x+1}\right)^2=5^2$ **1c.** $4x+1=25$ **1d.** $x=6$ **1e.** yes **1f.** $\{6\}$

2. Extraneous solutions are apparent solutions; they are algebraically correct but do not satisfy the original equation. **3.** Isolate the radical on one side of the equation. If there are two radicals, one radical should be on the left side and the other radical on the right side of the equation. **4a.** $x=1$ **4b.** no **4c.** $\varnothing$ **4d.** The square root function is equal to a negative number. **5a.** $x=4$ or $x=-1$ **5b.** $x=4$, yes ; $x=-1$, no

5c. $\{4\}$ **6a.** $(3x+1)^{\frac{3}{2}}=8$ **6b.** $\dfrac{2}{3}$ **6c.** $x=1$ **6d.** $\{1\}$ **7a.** $\sqrt{2x^2-5x-20}=\sqrt{x^2-3x+15}$

7b. $2x^2-5x-20=x^2-3x+15$ **7c.** $x=7,\ x=-5$ **7d.** yes **7e.** $\{7,-5\}$

Do the Math **1.** $\{14\}$ **2.** $\varnothing$ **3.** $\{3\}$ **4.** $\{49\}$ **5.** $\{13\}$ **6.** $\{9\}$ **7.** $\{4\}$ **8.** $\{-2\}$ **9.** $\{6\}$ **10.** $\{-6,3\}$

11. $\{1,5\}$ **12.** $\{12\}$ **13.** $\{-6\}$ **14.** $\{3\}$ **15.** $\{5\}$ **16.** $\{9\}$ **17.** $\{12\}$ **18.** $\{9\}$ **19.** $a=\dfrac{v^2}{r}$ **20.** $S=4\pi r^2$

Do the Math **21.** $U = \dfrac{CV^2}{2}$ **22.** $1102.50

Section 6.8

Five-Minute Warm-Up **1a.** $\{8, |-5|\}$ **1b.** $\left\{8, |-5|, \dfrac{0}{-12}\right\}$ **1c.** $\left\{8, |-5|, \dfrac{0}{-12}, \dfrac{-6}{3}\right\}$

1d. $\left\{8, |-5|, \dfrac{0}{-12}, \dfrac{-6}{3}, 0.\overline{3}, -\dfrac{2}{5}\right\}$ **1e.** $\{\pi\}$ **1f.** $\left\{8, |-5|, \dfrac{0}{-12}, \dfrac{-6}{3}, 0.\overline{3}, -\dfrac{2}{5}, \pi\right\}$ **2.** $3x^2 - 2$

3. $-12p^6 + 36p^5$ **4.** $-6x^2 + 13x + 28$ **5.** $x^4 - 16$

Guided Practice **1.** $-1;\ \sqrt{-1}$ **2.** $a + bi$; real numbers; standard; 6; -2 **3.** $9i$ **4.** $6 + 2i$ **5.** standard **6.** Add the real parts, then add the imaginary parts together. This is similar to combining like terms for polynomials. **7.** $12 + 9i$ **8a.** It cannot be used because $\sqrt{-4}$ and $\sqrt{-9}$ are not real numbers. **8b.** $2i \bullet 3i$

8c. -6 **9a.** $4 - 4i$ **9b.** $\dfrac{3 + 6i}{4 + 4i} \cdot \dfrac{4 - 4i}{4 - 4i}$ **9c.** $\dfrac{12 - 12i + 24i - 24i^2}{16 + 16}$ **9d.** $\dfrac{36 + 12i}{32}$ **9e.** $\dfrac{36}{32} + \dfrac{12i}{32}$

9f. $\dfrac{9}{8} + \dfrac{3}{8}i$ **10a.** -1 **10b.** $-i$ **10c.** 1 **10d.** i **10e.** -1 **10f.** $-i$ **10g.** 1

Do the Math **1.** $-10i$ **2.** $9\sqrt{2}\,i$ **3.** $10 + 4\sqrt{2}\,i$ **4.** $2 - i$ **5.** $3 - \sqrt{2}\,i$ **6.** $-3 + 14i$ **7.** $-4 + i$ **8.** $-3 + i$

9. $-4 - \sqrt{5}\,i$ **10.** $18 - 6i$ **11.** $5 + 5i$ **12.** $2 - 26i$ **13.** $\dfrac{5}{3} + \dfrac{5}{3}i$ **14.** $-21 + 20i$ **15.** $-45 - 28i$ **16.** -12

17. $54 + 23i$ **18.** $-\dfrac{1}{2} - i$ **19.** $\dfrac{8}{17} - \dfrac{2}{17}i$ **20.** $\dfrac{10}{17} - \dfrac{6}{17}i$ **21.** i **22.** 1 **23.** -1 **24.** $-i$

Chapter 7 Answers

Section 7.1

Five-Minute Warm-Up **1.** $9x^2 - 6x + 1$ **2.** $(x - 2)^2$ **3.** $\left\{\dfrac{1}{3}\right\}$ **4.** $\left\{\dfrac{7}{5}, -\dfrac{7}{5}\right\}$ **5.** $\dfrac{9}{5}$ **6.** $|8x - 3|$

7. $2\sqrt{3}\,i$ **8.** $2 + \dfrac{1}{2}i$ **9.** $-15 - 7i$

Guided Practice **1.** $x = \sqrt{p}$ or $x = -\sqrt{p}$ $\left(x = \pm\sqrt{p}\right)$ **2.** $\left\{-1 + 2\sqrt{3}, -1 - 2\sqrt{3}\right\}$ **3.** False **4.** False

5a. $n^2 = 144$ **5b.** $\sqrt{n^2} = \pm\sqrt{144}$ **5c.** $n = \pm 12$ **5d.** $\{12, -12\}$ **6a.** integer, rational, irrational

6b. complex; $a + bi, b \neq 0$ **7.** $\left(\dfrac{1}{2}b\right)^2$ **8a.** 49; $(p - 7)^2$ **8b.** $\dfrac{81}{4}; \left(n + \dfrac{9}{2}\right)^2$ **8c.** $\dfrac{4}{9}; \left(z + \dfrac{2}{3}\right)^2$

9a. $p^2 - 6p = 18$ **9b.** 9 **9c.** $p^2 - 6p + 9 = 27$ **9d.** $(p - 3)^2 = 27$ **9e.** $\sqrt{(p - 3)^2} = \pm\sqrt{27}$

9f. $p - 3 = \pm 3\sqrt{3}$ **9g.** $p = 3 \pm 3\sqrt{3}$ **9h.** $\left\{3 + 3\sqrt{3}, 3 - 3\sqrt{3}\right\}$ **10.** divide both sides of the equation by 3.

11. In a right triangle, the square of the length of the hypotenuse is equal to the sum of the squares of the lengths of the legs; $z^2 = x^2 + y^2$

Do the Math **1.** $\left\{-4\sqrt{3}, 4\sqrt{3}\right\}$ **2.** $\left\{-2\sqrt{5}, 2\sqrt{5}\right\}$ **3.** $\{-1, 5\}$ **4.** $\left\{-\dfrac{7}{2}, \dfrac{1}{2}\right\}$ **5.** $\left\{-\dfrac{3}{2} \pm \dfrac{\sqrt{3}}{2}\right\}$

6. $\{-1, 7\}$ **7.** $4; (p - 2)^2$ **8.** $\dfrac{1}{36}; \left(z - \dfrac{1}{6}\right)^2$ **9.** $\dfrac{25}{16}; \left(m + \dfrac{5}{4}\right)^2$ **10.** $\{-6, 3\}$ **11.** $\left\{-\dfrac{7}{2} \pm \dfrac{\sqrt{21}}{2}\right\}$

12. $\left\{\dfrac{5}{2} \pm \dfrac{\sqrt{37}}{2}\right\}$ **13.** $\left\{5 \pm \sqrt{30}\right\}$ **14.** $\left\{-\dfrac{2}{3}, 2\right\}$ **15.** $\left\{-\dfrac{3}{2} \pm \dfrac{1}{2}i\right\}$ **16.** 25 **17.** 3 **18.** $4\sqrt{6} \approx 9.80$

19. $-1 \pm 4\sqrt{2}$ **20.** ≈ 82.284 feet **21.** right triangle; hypotenuse is 52

Section 7.2

Five-Minute Warm-Up   **1.** $5\sqrt{5}$ **2.** $\dfrac{5}{2}+\sqrt{2}$ **3.** $2\sqrt{7}i$ **4.** $1-\dfrac{2}{3}i$ **5.** $3x^2-9x+1$ **6.** $2\sqrt{10}$

Guided Practice   **1.** $\dfrac{-b\pm\sqrt{b^2-4ac}}{2a}$ **2.** standard form   **3a.** $2x^2-x-4=0$; $a=2,\ b=-1,\ c=-4$

3b. $3x^2+6=0$; $a=3,\ b=0,\ c=6$ **3c.** $3x^2-6x=0$; $a=3,\ b=-6,\ c=0$ **4a.** 8; -2; -3

4b. $n=\dfrac{-b\pm\sqrt{b^2-4ac}}{2a}$ **4c.** $n=\dfrac{-(-2)\pm\sqrt{(-2)^2-4(8)(-3)}}{2(8)}$ **4d.** 100 **4e.** $\dfrac{2\pm10}{16}=\dfrac{1\pm5}{8}$ **4f.** $\dfrac{1+5}{8},\ \dfrac{1-5}{8}$

4g. $\left\{\dfrac{3}{4},-\dfrac{1}{2}\right\}$ **5.** b^2-4ac **6a.** 169; two rational   **6b.** -20; two complex, not real   **6c.** 0; one repeated real

7a. 200 **7b.** $200=-16t^2+150t+2$ **7c.** $\{1.6,\ 7.8\}$ **7d.** The rocket will be at a height of 200 feet at two different times. While it is going up, 1.6 seconds after launch, and again when it comes down, at 7.8 seconds. **7e.** no   **7f.** At 9.4 seconds the rocket will hit the ground ($h=0$).

Do the Math   **1.** $\{-4,8\}$ **2.** $\left\{-\dfrac{1}{2},\dfrac{2}{5}\right\}$ **3.** $\left\{1\pm\dfrac{\sqrt{2}}{2}\right\}$ **4.** $\left\{\dfrac{3}{2}\pm\dfrac{\sqrt{5}}{2}\right\}$ **5.** $\left\{1\pm\dfrac{\sqrt{10}}{2}i\right\}$ **6.** $\left\{-\dfrac{3}{5}\pm\dfrac{\sqrt{14}}{5}\right\}$

7. 24, 2 irrational   **8.** 0, one repeated real   **9.** -95, 2 complex, not real   **10.** $\left\{\dfrac{7}{2}\pm\dfrac{\sqrt{21}}{2}\right\}$ **11.** $\left\{-2,\dfrac{1}{3}\right\}$

12. $\left\{\dfrac{2}{5}\pm\dfrac{\sqrt{29}}{5}\right\}$ **13.** $\left\{\dfrac{5}{2}\right\}$ **14.** $\{1\pm\sqrt{6}\}$ **15.** $\left\{\dfrac{1}{2}\pm\dfrac{\sqrt{19}}{2}i\right\}$ **16a.** $x=-4$ or $x=2$ **16b.** $(-2,-8);\ (0,-8)$

17. width ≈5.307 in.; length ≈11.307 in.   **18.** base ≈9.426 in.; height ≈7.426 in.   **19.** ≈49.0 mph

20. Let $x_1=\dfrac{-b+\sqrt{b^2-4ac}}{2a}$; $x_2=\dfrac{-b-\sqrt{b^2-4ac}}{2a}$. $x_1+x_2=\dfrac{-2b}{2a}=-\dfrac{b}{a}$; $x_1\cdot x_2=\dfrac{b^2-(b^2-4ac)}{2a\cdot2a}=\dfrac{4ac}{4a^2}=\dfrac{c}{a}$

Section 7.3

Five-Minute Warm-Up   **1.** $(a-3)(a+3)(a^2+2)$ **2.** $(2x-5)(3x-5)$ **3.** $\dfrac{25}{x^2}$ **4.** $\dfrac{4}{9}x^6$ **5.** $\left\{\dfrac{3}{2},-2\right\}$

Guided Practice   **1a.** $\sqrt{y}$; $2u^2-11u+15=0$ **1b.** $\dfrac{1}{v^2}$; $u^2+10u+1=0$ **1c.** $(x-1)$; $u^2-5u+6=0$

1d. $x^{\frac{1}{3}}$; $u^2-u-3=0$ **1e.** x^{-1} or $\dfrac{1}{x}$; $4u^2+u-3=0$ **2a.** x^2 **2b.** $u^2-6u-16=0$ **2c.** $(u-8)(u+2)=0$

2d. $u=8$ or $u=-2$ **2e.** $x^2=8$ or $x^2=-2$ **2f.** $\sqrt{x^2}=\pm\sqrt{8}$ or $\sqrt{x^2}=\pm\sqrt{-2}$

2g. $x=\pm2\sqrt{2}$ or $x=\pm\sqrt{2}i$ **2h.** $\{2\sqrt{2},-2\sqrt{2},\sqrt{2}i,-\sqrt{2}i\}$ **3a.** (x^2+3) **3b.** $\{1,-1,i,-i\}$ **4a.** $\sqrt{x}$

4b. $\{4\}$ **5a.** $n^{\frac{1}{3}}$ **5b.** $\{-1,27\}$ **6a.** For $\sqrt{x}$, $x\ge0$.   **6b.** For $\dfrac{1}{x}$, $x\ne0$.

Do the Math   **1.** $\{-3,-1,1,3\}$ **2.** $\left\{-1,-\dfrac{1}{2},\dfrac{1}{2},1\right\}$ **3.** $\{3,-4\}$ **4.** $\{36\}$ **5.** $\varnothing$ **6.** $\left\{-\dfrac{1}{5},\dfrac{1}{3}\right\}$

7. $\left\{-\dfrac{2}{5},5\right\}$ **8.** $\{-1,27\}$ **9.** $\left\{\dfrac{1}{3},\dfrac{1}{4}\right\}$ **10.** $\left\{-1,2,-1\pm\sqrt{3}i,\dfrac{1}{2}\pm\dfrac{\sqrt{3}}{2}i\right\}$ **11.** $\{2,-3\}$ **12.** $\{-2i,-1,1,2i\}$

13. $\{81\}$ **14.** $\left\{\dfrac{7}{8},2\right\}$ **15.** $0,\sqrt{5}i,-\sqrt{5}i$ **16.** $-\sqrt{7}i,\sqrt{7}i,-\sqrt{2},\sqrt{2}$ **17.** $0,-\sqrt{3},\sqrt{3}$

18. $-\sqrt{2}i,\sqrt{2}i,-\sqrt{5},\sqrt{5}$ **19.** $-\sqrt{6},\sqrt{6},-\sqrt{7},\sqrt{7}$ **20.** $\dfrac{49}{4}$

Sullivan/Struve, *Intermediate Algebra*, 3e
Copyright © 2014 Pearson Education, Inc.

Section 7.4

Five-Minute Warm-Up   1 – 2. See Graphing Answer Section **3.** -29 **4.** 33 **5.** $\{x \mid x \text{ is any real number}\}$

Guided Practice   1. (b)   **2.** (a)   **3a.** vertically   **3b.** 2 **3c.** down   **4a.** horizontally   **4b.** 1 **4c.** right   **5a.** up
5b. down   **6a.** >1 **6b.** $0 < |a| < 1$ **7a.** $f(x) = \left(3x^2 - 12x\right) + 7$ **7b.** $f(x) = 3\left(x^2 - 4x + \underline{}\right) + 7 + \underline{}$

7c. 4 **7d.** Add the additive inverse of $(3)(4)$ or -12 **7e.** $f(x) = 3\left(x^2 - 4x + 4\right) + 7 + (-12)$

7f. $f(x) = 3(x - 2)^2 - 5$ **7g.** $(2, -5)$ **7h.** up   **7i.** $x = 2$

Do the Math   1 – 6. See Graphing Answer Section   **7.** $h(x) = \left(x - \dfrac{7}{2}\right)^2 - \dfrac{9}{4}; \; V\left(\dfrac{7}{2}, -\dfrac{9}{4}\right); \; x = \dfrac{7}{2}$

8. $f(x) = 3(x + 3)^2 - 2; \; V(-3, -2); \; x = -3$ **9.** $g(x) = -(x + 4)^2 + 2; \; V(-4, 2); \; x = -4$

10. $h(x) = -4\left(x - \dfrac{1}{2}\right)^2 + 1; \; V\left(\dfrac{1}{2}, 1\right); \; x = \dfrac{1}{2}$ **11.** $f(x) = \dfrac{1}{2}(x + 5)^2$ **12.** $f(x) = x^2 - 5$

Section 7.5

Five-Minute Warm-Up   1. x-intercept: $(-4, 0)$; y-intercept: $(0, 3)$ **2.** $\left\{8, -\dfrac{3}{2}\right\}$ **3.** $(6, 0)$ and $(-1, 0)$ **4.** -12

Guided Practice   1. $-\dfrac{b}{2a}$ **2.** setting $f(x) = 0$ and solving for x.   **3a.** $1; 2; -8$ **3b.** up   **3c.** $x = -1$

3d. $y = -9$ **3e.** $(-1, -9)$ **3f.** $x = -1$ **3g.** -8 **3h.** 36 **3i.** 2 **3j.** $(-4, 0); (2, 0)$ **3k.** See Graphing

Answer Section   **4.** y-coordinate   **5.** maximum   **6.** minimum **7a.** $x \cdot y$ **7b.** $2x + y = 50$

7c. $y = 50 - 2x$ **7d.** $A = x(50 - 2x)$ or $A = -2x^2 + 50x$ **7e.** $x = 12.5$ **7f.** $y = 25$

7g. The pen has two sides of 12.5 ft and one side 25 ft.   **7h.** 312.5 ft^2

Do the Math   1. $2; \left(-\dfrac{1}{2}, 0\right)$ and $(4, 0)$ **2.** $1; (3, 0)$ **3.** $2; (-0.28, 0)$ and $(1.78, 0)$ **4 – 9.** See Graphing

Answer Section   **10.** maximum; 11   **11.** minimum; -6 **12.** 25 and 25   **13.** -5 and 5   **14a.** ≈ 4.84 seconds
14b. ≈ 383.39 feet   **14c.** ≈ 9.74 seconds

Section 7.6

Five-Minute Warm-Up   1. $[-4, -2)$ **2.** $(-3, 1]$ **3.** $(-2, \infty)$ **4.** $(-\infty, 5]$ **5.** $\{x \mid x \geq 2\}$

6. $\left\{1 + \sqrt{6}, 1 - \sqrt{6}\right\}$

Guided Practice   1a. above   **1b.** below   **2a.** graphical   **2b.** algebraic   **3a.** $f(x) = x^2 + 2x - 3$
3b. $(-3, 0); (1, 0)$ **3c.** $(-1, -4)$ **3d.** See Graphing Answer Section   **3e.** negative   **3f.** $[-3, 1]$
4a. positive   **4b.** negative   **5a.** $(x + 7)(x - 5) = 0$ **5b.** $x = -7; x = 5$ **5c.** $(-\infty, -7); (-7, 5); (5, \infty)$
5d. See Graphing Answer Section   **5e.** positive   **5f.** no   **5g.** $(-\infty, -7) \cup (5, \infty)$

Do the Math   1. $[-1, 8]$ **2.** $(-\infty, 4) \cup (10, \infty)$ **3.** $(-4, -1)$ **4.** $\left(-\dfrac{7}{2}, 1\right)$ **5.** $(-\infty, -2) \cup (3, \infty)$

6. $\left(-\infty, \dfrac{3 - \sqrt{29}}{2}\right] \cup \left[\dfrac{3 + \sqrt{29}}{2}, \infty\right)$ **7.** $\left(-\infty, \dfrac{-3 - \sqrt{69}}{6}\right) \cup \left(\dfrac{-3 + \sqrt{69}}{6}, \infty\right)$ **8.** $(-\infty, \infty)$ **9.** $\varnothing$ **10.** $\{4\}$

11. $(-\infty, -4) \cup (0, \infty)$ **12.** $[-8, 6]$ **13.** $(-\infty, 0] \cup [5, \infty)$ **14.** $(-\infty, -9] \cup [7, \infty)$ **15.** between $50 and $70

16. $\left[-\dfrac{4}{3}, 2\right] \cup [6, \infty)$ **17.** $(-\infty, -2) \cup \left(-\dfrac{5}{3}, 2\right)$

Chapter 8 Answers
Section 8.1

Five-Minute Warm-Up **1.** $\{x \mid x \neq 2, x \neq 1\}$ **2a.** -47 **2b.** $-3a^2 + 12a - 11$

2c. $-3x^2 - 6xh - 3h^2 + 1$ **3a.** not a function **3b.** is a function

Guided Practice **1.** $(f \circ g)(x)$ **2a.** -3 **2b.** 25 **2c.** 25 **3a.** 39 **3b.** 38 **3c.** 38

4. A function is one-to-one if each unique element in the domain corresponds to a unique element in the range and each unique element in the range corresponds to a unique element in the domain. **5a.** yes **5b.** no **6.** the horizontal line test

7. one-to-one **8.** $f^{-1}(x)$; (b, a) **9a.** $\{(15, -3), (-5, -1), (0, 0), (10, 2)\}$ **9b.** $\{-3, -1, 0, 2\}$

9c. $\{15, -5, 0, 10\}$ **9d.** $\{15, -5, 0, 10\}$ **9e.** $\{-3, -1, 0, 2\}$ **10.** $y = x$ **11a.** $y = 4x - 8$

11b. $x = 4y - 8$ **11c.** $x + 8 = 4y$ **11d.** $y = \dfrac{x+8}{4}$ **11e.** $f^{-1}(x) = \dfrac{x+8}{4}$

Do the Math **1.** 253 **2.** 6 **3.** 1 **4.** -94 **5.** $(f \circ g)(x) = 4x - 3$ **6.** $(g \circ f)(x) = \sqrt{x+2} - 2$

7. $(f \circ f)(x) = \dfrac{2(x-1)}{3-x}$; $x \neq 1, 3$ **8.** one-to-one **9.** not one-to-one **10.** one-to-one

11. $\{(1, -10), (4, -5), (3, 0), (2, -5)\}$ **12.** $(f \circ g)(x) = (g \circ f)(x) = x$ **13.** $(f \circ g)(x) = (g \circ f)(x) = x$

14. $g^{-1}(x) = x - 6$ **15.** $H^{-1}(x) = \dfrac{x-8}{3}$ **16.** $f^{-1}(x) = \sqrt[3]{x+2}$ **17.** $G^{-1}(x) = 3 - \dfrac{2}{x}$

18. $R^{-1}(x) = \dfrac{4x}{2-x}$ **19.** $g^{-1}(x) = (x+3)^3 - 2$ **20.** $V(t) = 36\pi t$; $1080\pi \approx 3392.92 \, m^3$

Section 8.2

Five-Minute Warm-Up **1.** 16 **2.** $\dfrac{1}{4}$ **3.** 1 **4.** $\dfrac{1}{10}$ **5a.** 6.0235 **5b.** 6.0234 **6.** $\dfrac{10}{x^2}$ **7.** $\dfrac{2a^6}{5}$

8. $64p^{15}$ **9.** $\left\{-\dfrac{1}{2}, \dfrac{5}{3}\right\}$

Guided Practice **1.** a positive real number; 1 **2.** ≈ 2.63902 **3.** ≈ 2.66514 **4.** (b) **5.** (a) **6.** ≈ 2.718

7a. ≈ 20.09 **7b.** ≈ 0.14 **8.** the exponents are equal **9a.** $3^{2x+1} = 3^3$ **9b.** $2x + 1 = 3$ **9c.** $2x = 2$

9d. $x = 1$ **9e.** $\{1\}$ **10.** ≈ 64.7 grams **11.** $A = P\left(1 + \dfrac{r}{n}\right)^{nt}$; A = final amount; t = years that the interest

accumulates; r = annual interest rate written in decimals; n = number of pay periods per year
12a. $P=\$100$, $r= .07$, $n=4$, $t=5$ **12b.** \$141.48 **12c.** $P=\$100$, $r= .07$, $n=360$, $t=5$ **12d.** \$141.90
Do the Math **1a.** 9.518 **1b.** 9.673 **1c.** 9.735 **1d.** 9.738 **1e.** 9.739 **2a.** 20.086 **2b.** 4.482

3 – 5. See Graphing Answer Section **6.** $\{-2\}$ **7.** $\{4\}$ **8.** $\{1\}$ **9.** $\left\{-1, -\dfrac{1}{3}\right\}$ **10.** $\{-2\}$ **11.** $\left\{\dfrac{2}{3}\right\}$

12. $f(2) = 9$; $(2, 9)$ **13.** $x = -4$; $\left(-4, \dfrac{1}{81}\right)$ **14a.** ≈ 7287 million people **14b.** ≈ 8259 million people

Section 8.3

Five-Minute Warm-Up **1.** $\left\{x \mid x > -\dfrac{3}{2}\right\}$ **2.** $\{3\}$ **3.** $\{-4, -3\}$ **4.** $\dfrac{1}{16}$ **5.** 8 **6.** 4 **7.** $\dfrac{1}{4}$

Guided Practice **1.** $x = a^y$ **2.** $\log_2\left(\dfrac{1}{8}\right) = -3$ **3.** $\log_a 3 = 4$ **4.** $4^p = 30$ **5.** $n^{-5} = 3$ **6a.** $(0, \infty)$

6b. $(-\infty, \infty)$ **6c.** $\{x \mid x < 5\}$ **7.** $y = 5^x$; See Graphing Answer Section **8.** $x = e^y$ **9.** $x = 10^y$ **10.** $\{-2\}$

11. $e^{-2} \approx 0.1353$ **12a.** decibels (loudness) **12b.** Richter Scale (earthquakes) **12c.** pH (chemistry)

Sullivan/Struve, *Intermediate Algebra*, 3e
Copyright © 2014 Pearson Education, Inc.

Do the Math **1.** $\log_4 64 = 3$ **2.** $\log_b 23 = 4$ **3.** $\log z = -3$ **4.** $3^4 = 81$ **5.** $6^{-4} = x$ **6.** $a^2 = 16$ **7.** 2

8. 2 **9.** $(2, \infty)$ **10.** $\left(-\infty, \dfrac{3}{5}\right)$ **11.** -0.108 **12.** -0.693 **13.** $\{6\}$ **14.** $\left\{\dfrac{27}{4}\right\}$ **15.** $\{9\}$ **16.** $\left\{\dfrac{7}{2}\right\}$ **17.** $\{4\}$

18. $\left\{-2\sqrt{2}, 2\sqrt{2}\right\}$ **19.** ≈ 9.2 on the Richter scale

Section 8.4

Five-Minute Warm-Up **1a.** 1.140 **1b.** 1.139 **2.** $x^{\frac{1}{2}}$ **3.** $a^{\frac{3}{4}}$ **4.** 2^{-4} **5.** $\left(\dfrac{3}{2}\right)^3$ **6.** 1 **7.** 6 **8.** $y = 3^x$

Guided Practice **1a.** $\sqrt{2}$ **1b.** 0 **1c.** 1 **1d.** 7 **2a.** $\log_a M + \log_a N$ **2b.** $\log_a M - \log_a N$

2c. $r \log_a M$ **3.** $\log 9 + \log x$ **4.** $\ln 2 - \ln x$ **5.** $\log_4\left(\dfrac{1}{x-1}\right)$ **6.** $\dfrac{\log_b M}{\log_b a}$ **7.** ≈ 3.170 **8.** Answers may vary.

Do the Math **1a.** -3 **1b.** $\sqrt{2}$ **1c.** 10 **2a.** $2a$ **2b.** $a + 2b$ **3.** $\log_4 a - \log_4 b$ **4.** $3 \log_3 a + \log_3 b$

5. $3 + \log_2 z$ **6.** $4 - \log_2 p$ **7.** $5 + \dfrac{1}{4}\log_2 z$ **8.** $\dfrac{1}{5}\ln x - 2\ln(x+2)$ **9.** $\dfrac{1}{3}\left[\log_6(x-2) - \log_6(x+1)\right]$

10. $3\log_4 x + \log_4(x-3) - \dfrac{1}{3}\log_4(x+1)$ **11.** 3 **12.** 4 **13.** $\log_2 z^8$ **14.** $\log_2\left(a^4 b^2\right)$

15. $\log_4\left[\sqrt[3]{z}\,(2z+1)^2\right]$ **16.** $\log_7\left(x^2\right)$ **17.** $\ln\sqrt[3]{x^2 - 1}$ **18.** $\log_5(x+1)$ **19.** $\log_4\left(\dfrac{x^4}{16}\right)$ **20a.** 0.827

20b. 1.631

Section 8.5

Five-Minute Warm-Up **1.** $\{9\}$ **2.** $\{-8, 3\}$ **3.** $\left\{\dfrac{1}{4}, -2\right\}$ **4.** $\{7, 0\}$ **5.** $\{x \mid x < 3\}$

Guided Practice **1a.** $\left\{\dfrac{5}{2}\right\}$ **1b.** $\left\{\dfrac{1}{3}\right\}$ **1c.** $\{64\}$ **2a.** $x = \dfrac{\log 12}{\log 3} \approx 2.262$ **2b.** $x = \dfrac{\ln 18}{3} \approx 0.963$

3a. $90 = 100\left(\dfrac{1}{2}\right)^{\frac{t}{19.255}}$ **3b.** ≈ 2.927 sec. **4a.** 3921 people **4b.** $7500 = 2500e^{0.03t}$ **4c.** 36.6 years or 1986

4d. 23.1 years or 1973

Do the Math **1.** $\{13\}$ **2.** $\{16\}$ **3.** $\{8\}$ **4.** $\left\{\dfrac{5}{8}\right\}$ **5.** $\{5\}$ **6.** $\dfrac{\log 8}{\log 3} \approx 1.893$ **7.** $\dfrac{\log 20}{\log 4} \approx 2.161$

8. $\ln 3 \approx 1.099$ **9.** $\log 0.2 \approx -0.699$ **10.** $\dfrac{\log 5}{2\log 2} \approx 1.161$ **11.** $\dfrac{\log 5}{\log 4} \approx 1.161$ **12.** $\{4\}$ **13.** $\{2\}$ **14.** $\{2\}$

15. $\ln 4 \approx 1.386$ **16.** $\{12\}$ **17.** $\left\{\pm 2\sqrt{2} \approx -2.828, 2.828\right\}$ **18.** $\{6\}$ **19.** 2052 **20.** ≈ 2.989 years

Chapter 9 Answers
Section 9.1

Five-Minute Warm-Up **1.** $2\sqrt{10}$ **2.** $6\sqrt{3}$ **3.** $4p^2\sqrt{2}$ **4.** $|2x - 5|$ **5.** -10 **6.** 25 **7.** 6 cm^2 **8.** 12

Guided Practice **1.** $d = \sqrt{(x_2 - x_1)^2 + (y_2 - y_1)^2}$ **2a.** a and c **2b.** b and d **2c.** (a) **2d.** (b)

2e. add (c) and (d) **2f.** take the square root of (e) **2g.** simplify **3.** $4\sqrt{5} \approx 8.94$ **4.** two equal parts

5. $x_1 + x_2; y_1 + y_2$ **6.** $(-4, -3)$

Do the Math **1.** 5 **2.** 25 **3.** 2 **4.** $2\sqrt{13} \approx 7.21$ **5.** $2\sqrt{78} \approx 17.66$ **6.** $\sqrt{5.69} \approx 2.39$ **7.** $(3, 5)$ **8.** $(2, 2)$

9. $\left(1, \dfrac{11}{2}\right)$ **10.** $(2, -2)$ **11.** $\left(2\sqrt{6}, 4\sqrt{2}\right)$ **12.** $(-0.7, 1.95)$ **13a.** See Graphing Answer Section

13b. $d(A, B) = 5$; $d(B, C) = 3\sqrt{5}$; $d(A, C) = \sqrt{58} \approx 7.62$ **14.** $(-16, -3)$ and $(8, -3)$

Section 9.2

Five-Minute Warm-Up 1. $49; (x-7)^2$ 2. $\dfrac{25}{4}; \left(y-\dfrac{5}{2}\right)^2$ 3. $(y+8)^2$ 4. $2(x-3)^2$

5a. $\dfrac{225\pi}{4} \approx 176.71$ in.^{2} **5b.** $15\pi \approx 47.12$ in.

Guided Practice 1a. center 1b. radius 1c. $d=2r$ 2. $(x-h)^2 + (y-k)^2 = r^2$

3. $(x+1)^2 + (y-3)^2 = 25$ 4a. $(2,-3)$ 4b. 2 4c. See Graphing Answer Section

5. $x^2 + y^2 + ax + by + c = 0$ 6a. 9; 1; 9; 1 6b. 3; 1; 9 6c. $(-3,1)$; 3 6d. See Graphing Answer ·ction

7. No; all circles fail the vertical line test. 8. Yes; Center: (2, 0); radius = 3

Do the Math 1 – 8. See Graphing Answer Section 1. $x^2 + y^2 = 25$ 2. $(x-1)^2 + y^2 = 4$

3. $(x+4)^2 + (y-4)^2 = 16$ 4. $(x-5)^2 + (y-2)^2 = 7$ 5. $C(0,0), r=5$ 6. $C(5,-2), r=7$

7. $C(6,0), r=6$ 8. $C(2,-2), r=0.5$ 9. $C(-1,4), r=3$ 10. $C(-2,6), r=2$ 11. $C(7,-5), r=8$

12. $x^2 + (y-3)^2 = 25$ 13. $(x-2)^2 + (y+3)^2 = 9$ 14. $(x-1)^2 + \left(y+\dfrac{1}{2}\right)^2 = \dfrac{169}{4}$

15. $A = 49\pi$ square units; $C = 14\pi$ units

Section 9.3

Five-Minute Warm-Up 1. $(2,1)$ 2. up 3. $x=2$ 4. $25; (x+5)^2$ 5. $-12; -3(x-2)^2$ 6. $\{8,0\}$

Guided Practice 1a. focus 1b. directrix 1c. vertex 1d. axis of symmetry 2a. left or right 2b. rig

2c. left 2d. up or down 2e. up 2f. down 3a. down 3b. right 4a. left 4b. $y^2 = -4ax$ 4c. $y^2 = -\dfrac{1}{2}$

5a. $x^2 - 8x = -4y - 20$ 5b. $x^2 - 8x + 16 = -4y - 20 + 16$ 5c. $x^2 - 8x + 16 = -4y - 4$

5d. $(x-4)^2 = -4(y+1)$ 5e. down 5f. $(4,-1)$ 5g. See Graphing Answer Section

Do the Math 1. $x^2 = 20y$ 2. $y^2 = -32x$ 3. $y^2 = 2x$ 4. $y^2 = 16x$ 5. $x^2 = -8y$

6 – 13. See Graphing Answer Section 6a. $V(0,0)$ 6b. $F(0,7)$ 6c. $y=-7$ 7a. $V(0,0)$, 7b. $F\left(\dfrac{5}{2}, 0\right)$

7c. $x = -\dfrac{5}{2}$ 8a. $V(0,0)$ 8b. $F(0,-4)$ 8c. $y=4$ 9a. $V(-4,1)$ 9b. $F(-4,0)$ 9c. $y=2$ 10a. $V(-5,2)$

10b. $F(-2,2)$ 10c. $x=-8$ 11a. $V(-1,3)$ 11b. $F(-1,5)$ 11c. $y=1$ 12a. $V(2,4)$ 12b. $F(-2,4)$

12c. $x=6$ 13a. $V(2,0)$ 13b. $F\left(2,-\dfrac{5}{2}\right)$ 13c. $y=\dfrac{5}{2}$ 14. 20 feet

Section 9.4

Five-Minute Warm-Up 1. $\dfrac{81}{4}; \left(y-\dfrac{9}{2}\right)^2$ 2. $36; (x+6)^2$ 3. $\dfrac{1}{4}; 4\left(x-\dfrac{1}{4}\right)^2$ 4. $y = 9(x+3)^2 - 6$

5. $(-3,-6)$ 6. up 7. $x=-3$

Guided Practice 1a. foci 1b. major axis 1c. minor axis 1d. center 1e. vertices 2a. a^2 2b. a

2c. $2a$ 2d. $2b$ 2e. major 2f. $a^2 - b^2$ 2g. the origin 3a – 3b. See Graphing Answer Section

4. $\dfrac{x^2}{144} + \dfrac{y^2}{169} = 1$ 5a. $\dfrac{(x+5)^2}{9} + \dfrac{(y-2)^2}{25} = 1$ 5b. $C(-5,2)$ 5c. $V_1(-5,7); V_2(-5,-3)$

5d. $F_1(-5,6); F_2(-5,-2)$ 6a. $\dfrac{(x-1)^2}{9} + \dfrac{(y-3)^2}{4} = 1$ 6b. $C(1,3)$ 6c. $V_1(4,3); V_2(-2,3)$ 6d. $F\left(1 \pm \sqrt{5}, 3\right)$

7a. $\dfrac{(x-4)^2}{36} + \dfrac{(y+3)^2}{64} = 1$ 7b. $C(4,-3)$ 7c. $V_1(4,-11); V_2(4,5)$ 7d. $F\left(4, -3 \pm 2\sqrt{7}\right)$

Do the Math 1 – 4. See Graphing Answer Section 1a. $V(\pm 5, 0)$ 1b. $F\left(\pm\sqrt{21}, 0\right)$ 2a. $V(0, \pm 6)$

Sullivan/Struve, *Intermediate Algebra*, 3e
Copyright © 2014 Pearson Education, Inc.

2b. $F\left(0, \pm 2\sqrt{5}\right)$ **3a.** $V(\pm 8, 0)$ **3b.** $F\left(\pm 3\sqrt{7}, 0\right)$ **4a.** $V(0, \pm 9)$ **4b.** $F\left(0, \pm 6\sqrt{2}\right)$ **5.** $\dfrac{x^2}{25} + \dfrac{y^2}{21} = 1$

6. $\dfrac{x^2}{24} + \dfrac{y^2}{25} = 1$ **7.** $\dfrac{x^2}{45} + \dfrac{y^2}{49} = 1$ **8.** $\dfrac{x^2}{100} + \dfrac{y^2}{64} = 1$ **9 – 10.** See Graphing Answer Section

11a. $\dfrac{(x+3)^2}{9} + \dfrac{(y-4)^2}{25} = 1$ **11b.** $C(-3, 4)$ **11c.** $V(-3, -1)$ and $(-3, 9)$ **11d.** $F(-3, 0)$ and $(-3, 8)$

12a. $\dfrac{x^2}{519.84} + \dfrac{y^2}{225} = 1$ **12b.** yes

Section 9.5

Five-Minute Warm-Up **1.** $\{-4, 4\}$ **2.** $\{-1, -5\}$ **3.** $16; (x+4)^2$ **4.** $\dfrac{4}{9}; \left(y - \dfrac{2}{3}\right)^2$ **5.** See Graphing

Answer Section
Guided Practice **1a.** foci **1b.** transverse axis **1c.** center **1d.** conjugate axis **1e.** vertices **2a.** a^2 **2b.** a
2c. $2a$ **2d.** $2b$ **2e.** left and right **2f.** up and down **2g.** $a^2 + b^2$ **2h.** the origin **3a.** center: (0,0) **3b.** a=4, b=3
3c. c=5 **3d.** vertices: $(\pm 4, 0)$; foci: $(\pm 5, 0)$ **3e.** $(\pm 5, 9), (\pm 5, -9)$ **3f.** See Graphing Answer Section

4. A boundary line that the graph approaches but does not cross as x (or y) $\to \pm\infty$. **5a.** $\dfrac{b}{a}x$ **5b.** $\dfrac{a}{b}x$

Do the Math **1 – 4.** See Graphing Answer Section **1.** $V(\pm 3, 0)$ **1b.** $F(\pm 5, 0)$ **2a.** $V(0, \pm 9)$

2b. $F\left(0, \pm 3\sqrt{10}\right)$ **3a.** $V(\pm 6, 0)$ **3b.** $F\left(\pm 2\sqrt{10}, 0\right)$ **4a.** $V(0, \pm 3)$ **4b.** $F\left(0, \pm\sqrt{13}\right)$ **5.** $x^2 - \dfrac{y^2}{15} = 1$

6. $\dfrac{y^2}{36} - \dfrac{x^2}{28} = 1$ **7.** $\dfrac{y^2}{16} - \dfrac{x^2}{4} = 1$ **8.** $\dfrac{x^2}{8.1} - \dfrac{y^2}{72.9} = 1$ **9.** parabola **10.** hyperbola **11.** hyperbola **12.** circle

13. parabola **14.** ellipse **15a.** (0, 5) and (0, –5) **15b.** $\left(0 \pm \sqrt{29}\right)$ **16.** $y = \pm\dfrac{3}{2}x$

Section 9.6

Five-Minute Warm-Up **1.** $\left(2, \dfrac{3}{2}\right)$ **2.** $\{(x, y) | -x + 3y = 4\}$ **3.** $(-4, -5)$ **4.** $\varnothing$

Guided Practice **1a.** a line **1b.** a circle **1c.** See Graphing Answer Section **1d.** 2 **1e.** $x^2 + y^2 = 16$
1f. $x^2 + (x-4)^2 = 16$ **1g.** $2x^2 - 8x = 0$ **1h.** $2x(x-4) = 0$ **1i.** $x = 0$ or $x = 4$ **1j.** $y = x - 4$
1k. $y = 0 - 4$ **1l.** $y = -4$ **1m.** $y = x - 4$ **1n.** $y = 4 - 4$ **1o.** $y = 0$ **1p.** $\{(0, -4), (4, 0)\}$

2a. a circle **2b.** an ellipse **2c.** See Graphing Answer Section **2d.** 2 **2e.** -1 **2f.** $\begin{cases} -x^2 - y^2 = -4 & (1) \\ x^2 + 4y^2 = 16 & (2) \end{cases}$

2g. $3y^2 = 12$ **2h.** $y^2 = 4$ **2i.** $y = \pm 2$ **2j.** $x^2 + y^2 = 4$ **2k.** $x^2 + (2)^2 = 4$ **2l.** $x = 0$

2m. $x^2 + y^2 = 4$ **2n.** $x^2 + (-2)^2 = 4$ **2o.** $x = 0$ **2p.** $\{(0, 2), (0, -2)\}$

Do the Math **1.** $(-1, 1)$, $(0, 2)$, and $(1, 3)$ **2.** $(6, 8)$ and $(8, 6)$ **3.** $(0, -4), \left(\pm\sqrt{7}, 3\right)$ **4.** $(1, 1)$

5. $(-2, -2)$ and $(2, -2)$ **6.** $(-2, 0)$ and $(2, 0)$ **7.** $(-3, 0)$ and $(3, 0)$ **8.** $(-2, -6)$ and $(1, 15)$ **9.** $\varnothing$
10. $(-5, 0)$ and $(5, 0)$ **11.** $\varnothing$ **12.** $(-4, -7), (4, -7), (-1, 8), (1, 8)$ **13.** –4 and 12 **14.** 20 meters by 12 meters

Chapter 10 Answers
Section 10.1

Five-Minute Warm-Up **1a.** -5 **1b.** -9 **2a.** 1 **2b.** 1 **2c.** -3 **2d.** 5 **3a.** $\dfrac{1}{9}$ **3b.** $-\dfrac{1}{27}$ **3c.** $\dfrac{1}{81}$ **4.** $\dfrac{25}{24}$

Guided Practice **1a.** terms **1b.** ellipsis **1c.** finite **2.** a_n **3.** $1, \dfrac{3}{2}, \dfrac{7}{3}, \dfrac{15}{4}, \dfrac{31}{5}$ **4a.** $a_n = 3n$

4b. $an = (-1)^{n+1}\left(\dfrac{1}{6}\right)^n$ **5.** 1; 2; 3; 4; 4; 7; 12; 19; 42   **6.** $-5 + (-3) + (-1) + 1 + 3 + 5 = 0$ **7.** $\displaystyle\sum_{i=0}^{6}(2i)$

Do the Math   **1.** $-3, -2, -1, 0, 1$ **2.** $5, 3, \dfrac{7}{3}, 2, \dfrac{9}{5}$ **3.** 2, 8, 26, 80, 242   **4.** $\dfrac{1}{2}, 2, \dfrac{9}{2}, 8, \dfrac{25}{2}$ **5.** $a_n = 5n$

6. $a_n = \dfrac{n}{2}$ **7.** $a_n = n^3 - 1$ **8.** $a_n = \left(-\dfrac{1}{2}\right)^{n-1}$ **9.** 55   **10.** 50   **11.** 120   **12.** 4   **13.** 16   **14.** 69

15. $\displaystyle\sum_{k=1}^{9}(2k-1)$ **16.** $\displaystyle\sum_{k=1}^{16}\dfrac{1}{2^{k-1}}$ **17.** $\displaystyle\sum_{i=1}^{15}(-1)^{i+1}\left(\dfrac{2}{3}\right)^i$ **18.** $\displaystyle\sum_{k=1}^{12}\left[3\cdot\left(\dfrac{1}{2}\right)^{k-1}\right]$ **19a.** \$5033.33
19b. \$5415.00   **19c.** \$11,098.20

Section 10.2

Five-Minute Warm-Up   **1.** 4   **2.** $-\dfrac{9}{5}$ **3a.** $(14, 4)$ **3b.** $(1, -4)$ **4.** -36

Guided Practice   **1a.** d **1b.** a_1 **2.** yes; $a_1 = 5$; $d = 3$ **3.** 3, 6, 11, 18, 27, 38; 3, 5, 7, 9, 11
4a. $a_n = 5n - 8$ **4b.** $a_8 = 32$ **5a.** $a_1 = 12$; $d = 3$ **5b.** $a_n = 3d + 12$ **6a.** $a_n = 4n + 8$ **6b.** $a_{10} = 48$
6c. $S_{10} = 300$ **7a.** $a_1 = 3$, $a_{10} = -87$ **7b.** $S_{10} = -420$

Do the Math   **1.** $d = 10$; 11, 21, 31, 41   **2.** $d = \dfrac{1}{4}$; $1, \dfrac{5}{4}, \dfrac{3}{2}, \dfrac{7}{4}$ **3.** $a_n = 3n + 5$; $a_5 = 20$

4. $a_n = -3n + 15$; $a_5 = 0$ **5.** $a_n = \dfrac{1}{2}n - \dfrac{7}{2}$; $a_5 = -1$ **6.** $a_n = 4n - 9$; $a_{20} = 71$ **7.** $a_n = -6n + 26$; $a_{20} = -94$

8. $a_n = -\dfrac{1}{2}n + \dfrac{21}{2}$; $a_{20} = \dfrac{1}{2}$ **9.** $a_n = 3n - 8$ **10.** $a_n = 4n - 17$ **11.** $a_n = -5n + 22$ **12.** $a_n = \dfrac{1}{4}n + \dfrac{15}{4}$
13. 5500   **14.** 10,425   **15.** -9200 **16.** 5440   **17.** -2905 **18.** -413 **19.** $x = 1$ **20.** 1600 seats

Section 10.3

Five-Minute Warm-Up   **1a.** $\dfrac{2}{3}$ **1b.** $\dfrac{4}{9}$ **1c.** $\dfrac{8}{27}$ **2a.** 3   **2b.** 12   **2c.** 27   **3a.** $\dfrac{8x^3}{5}$ **3b.** $9r^8$ **4.** 2

Guided Practice   **1a.** r **1b.** a_1 **2.** yes; $a_1 = 36$; $r = \dfrac{1}{2}$ **3.** $1, \dfrac{2}{5}, \dfrac{4}{25}, \dfrac{8}{125}$; $r = \dfrac{2}{5}$ **4a.** $\{a_n\} = \left\{2\cdot\left(-\dfrac{1}{3}\right)^{n-1}\right\}$

4b. $-\dfrac{2}{2187}$ **5a.** $a_1 = 6$, $r = 4$ **5b.** 1,048,575   **6a.** $a_1 = 6$; $r = -\dfrac{1}{3}$ **6b.** $\dfrac{9}{2}$ **7.** \$99,658.27

Do the Math   **1.** $r = -2$; $-2, 4, -8, 16$ **2.** $r = 2$; $\dfrac{2}{3}, \dfrac{4}{3}, \dfrac{8}{3}, \dfrac{16}{3}$ **3.** $r = \dfrac{1}{6}, \dfrac{1}{3}, \dfrac{1}{18}, \dfrac{1}{108}, \dfrac{1}{648}$ **4.** arithmetic

5. neither   **6.** geometric   **7.** neither   **8.** $a_n = 30\cdot\left(\dfrac{1}{3}\right)^{n-1}$; $a_8 = \dfrac{10}{729}$ **9.** $a_n = (-4)^{n-1}$; $a_8 = -16,384$

10. 177,147   **11.** -1280 **12.** 0.0000000004   **13.** 88,572   **14.** 40,950   **15.** $\dfrac{3}{2}$ **16.** $\dfrac{48}{5}$ **17.** 5   **18.** $\dfrac{1}{3}$

19. $\dfrac{5}{11}$ **20.** \$10,497.60

Section 10.4

Five-Minute Warm-Up   **1.** $x^2 + 4x + 4$ **2.** $y^2 - 6y + 9$ **3.** $16x^2 - 40xy + 25y^2$

4. $9n^2 + 4n + \dfrac{4}{9}$ **5.** 35

Guided Practice   **1a.** 151,200   **1b.** 15   **2a.** 10   **2b.** 1716   **3.** $p^4 + 8p^3 + 24p^2 + 32p + 16$
Do the Math   **1.** 120   **2.** 360   **3.** 45   **4.** 1   **5.** 10   **6.** 21   **7.** 50   **8.** 1   **9.** $x^4 - 4x^3 + 6x^2 - 4x + 1$
10. $x^5 + 25x^4 + 250x^3 + 1250x^2 + 3125x + 3125$ **11.** $16q^4 + 96q^3 + 216q^2 + 216q + 81$

12. $81w^4 - 432w^3 + 864w^2 - 768w + 256$ **13.** $y^8 - 12y^6 + 54y^4 - 108y^2 + 81$

14. $243b^{10} + 810b^8 + 1080b^6 + 720b^4 + 240b^2 + 32$ **15.** $p^6 - 18p^5 + 135p^4 - 540p^3 + 1215p^2 - 1458p + 729$

16. $81x^8 + 108x^6y^3 + 54x^4y^6 + 12x^2y^9 + y^{12}$ **17.** 1.00501

Graphing Answer Section

Section R.2 *Do the Math*

12.

Section R.3 Guided Practice

3.

	+	−
+	+	−
−	−	+

Section 1.4 Guided Practice

5g.

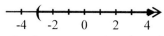

6h.

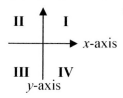

Section 1.4 *Do the Math*

1.

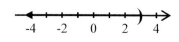

2.

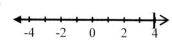

3.

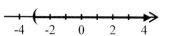

4.

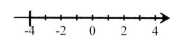

Section 1.5 Warm-Up

1.

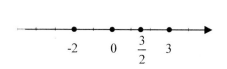

Section 1.5 Guided Practice

1.

II   I

x-axis

III   IV

y-axis

5.

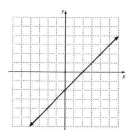

Section 1.5 *Do the Math*

2.

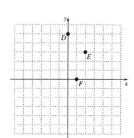

6.

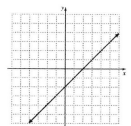

7.

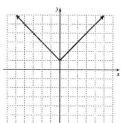

8.

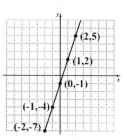

Section 1.5 *Do the Math*

9.

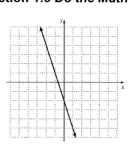

Section 1.6 Guided Practice

9m.

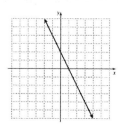

Section 1   Do the Math

1. 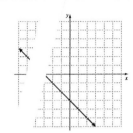   **2.** 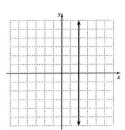   **3.**   **4.**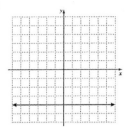

Se   n 1.6 *Do the Math*   Section 1.7 Guided Practice

7. 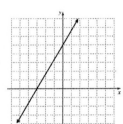   **4h.** 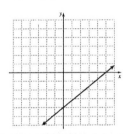   **8h.**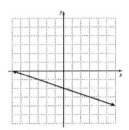

Section 1.8 Guided Practice   **Section 1.8 *Do the Math***

7h. 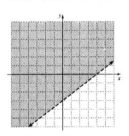   **3.** 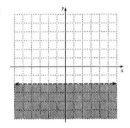   **4.**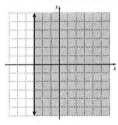

Section 1.8 *Do the Math*

5. 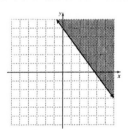   **6.** 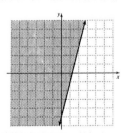   **7.** 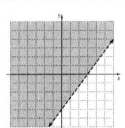   **8.**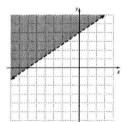

Section 2.1 Warm-Up

3. 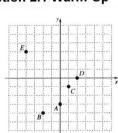   **4.** 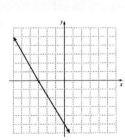   **5.**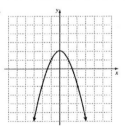

Sullivan/Struve, *Intermediate Algebra,* 3e
Copyright © 2014 Pearson Education, Inc

Section 2.1 Guided Practice

6.

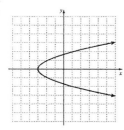

Section 2.3 Warm-Up

3.

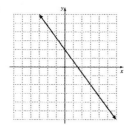

4.

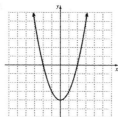

Section 2.3 Guided Practice

4.

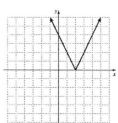

Section 2.3 *Do the Math*

6.

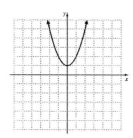

7.

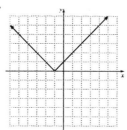

Section 2.4 Warm-Up

1.

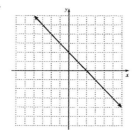

2.

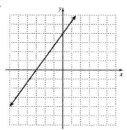

Section 2.4 Guided Practice

3c.

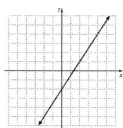

Section 2.4 *Do the Math*

1.

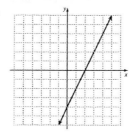

Section 2.4 *Do the Math*

2.

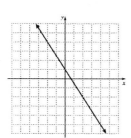

3.

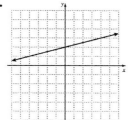

4.

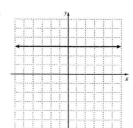

8.

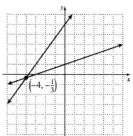

$\left(-4, -\frac{1}{3}\right)$

Section 2.5 Warm-Up

2.

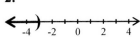

Section 2.5 Guided Practice

3a.

3b.

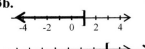

4d-f.

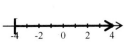

Section 2.5 Guided Practice

6d.

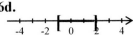

7e-g.

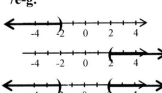

Section 2.6 Guided Practice

6d.

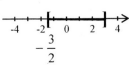

$$-\frac{3}{2}$$

8d.

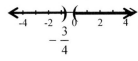

$$-\frac{3}{4}$$

Section 3.1 Warm-Up

3.

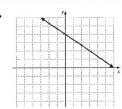

Section 3.4 Guided Practice

3.

$$\begin{bmatrix} 2 & 1 & 1 & | & 3 \\ 0 & 4 & -7 & | & -1 \\ 1 & 3 & 0 & | & 0 \end{bmatrix}$$

6a.

$$\begin{bmatrix} 2 & -1 & | & 5 \\ -2 & 14 & | & -8 \end{bmatrix}$$

6b.

$$\begin{bmatrix} 3 & -8 & | & 9 \\ 1 & -7 & | & 4 \end{bmatrix}$$

Section 3.4 Guided Practice

8a.

$$\begin{bmatrix} 1 & 1 & 1 & | & 1 \\ 2 & 2 & 0 & | & 6 \\ 3 & 4 & -1 & | & 13 \end{bmatrix}$$

8b.

$$\begin{bmatrix} 1 & 1 & 1 & | & 1 \\ 0 & 0 & -2 & | & 4 \\ 0 & 1 & -4 & | & 10 \end{bmatrix}$$

8c.

$$\begin{bmatrix} 1 & 1 & 1 & | & 1 \\ 0 & 1 & -4 & | & 10 \\ 0 & 0 & -2 & | & 4 \end{bmatrix}$$

8d.

$$\begin{bmatrix} 1 & 1 & 1 & | & 1 \\ 0 & 1 & -4 & | & 10 \\ 0 & 0 & 1 & | & -2 \end{bmatrix}$$

Section 3.4 *Do the Math*

1.

$$\begin{bmatrix} 6 & 4 & | & -2 \\ -1 & -1 & | & -1 \end{bmatrix}$$

2a.

$$\begin{bmatrix} 1 & -1 & 1 & | & 6 \\ 0 & -1 & -1 & | & 15 \\ 3 & 2 & -2 & | & -5 \end{bmatrix}$$

2b.

$$\begin{bmatrix} 1 & -1 & 1 & | & 6 \\ 0 & -1 & -1 & | & 15 \\ 0 & 5 & -5 & | & -23 \end{bmatrix}$$

Section 3.5 Guided Practice

6.

$$(1) \cdot \begin{vmatrix} -3 & 3 \\ 4 & -2 \end{vmatrix} - (-1) \cdot \begin{vmatrix} 5 & 3 \\ 1 & -2 \end{vmatrix} + (2) \cdot \begin{vmatrix} 5 & -3 \\ 1 & 4 \end{vmatrix}$$

Sullivan/Struve, *Intermediate Algebra*, 3e
Copyright © 2014 Pearson Education, Inc

Section 3.5 Guided Practice

7a.
$$\begin{vmatrix} 1 & 2 & -1 \\ 2 & -4 & 1 \\ -2 & 2 & -3 \end{vmatrix}$$

7c.
$$\begin{vmatrix} -3 & 2 & -1 \\ -7 & -4 & 1 \\ 4 & 2 & -3 \end{vmatrix}$$

7d.
$$\begin{vmatrix} 1 & -3 & -1 \\ 2 & -7 & 1 \\ -2 & 4 & -3 \end{vmatrix}$$

7e.
$$\begin{vmatrix} 1 & 2 & -3 \\ 2 & -4 & -7 \\ -2 & 2 & 4 \end{vmatrix}$$

Section 3.6 Warm-Up

4.

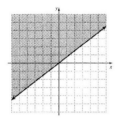

Section 3.6 Guided Practice

6.

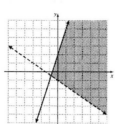

7d

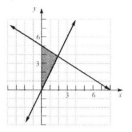

Do the Math 3.6

3.

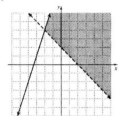

4.

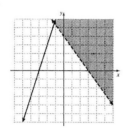

5.

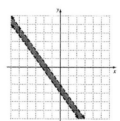

6.

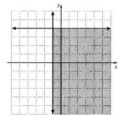

Do the Math 3.6

7.

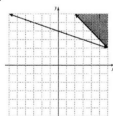

8a.
$$\begin{cases} 12x + 6y \le 1600 \\ 4x + 6y \le 1920 \\ x \ge 0 \\ y \ge 0 \end{cases}$$

Guided Practice 4.3

1c.

$$x + 1 \overline{\smash{\big)}\ x^3 - 2x^2 + x + 6} \quad \begin{array}{c} x^2 - 3x + 4 \end{array}$$

$$-\ \underline{\left(x^3 + x^2\right)}$$
$$-3x^2 + x + 6$$
$$-\ \underline{\left(-3x^2 - 3x\right)}$$
$$4x + 6$$
$$-\ \underline{\left(4x + 4\right)}$$
$$2$$

Warm-Up 5.5

2.

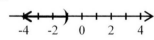

Guided Practice 4.5

7b.

Factors of –24	1, – 24	2, –12	3, – 8	4, – 6	6, –4	8, – 3	12, – 2	24, –1
Sum of 2	–23	–10	–5	–2	2	5	10	23

Guided Practice 5.5

5q.

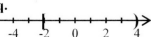

Guided Practice 5.5

8.

Interval	$(-\infty, -4)$	-4	$(-4, 2)$	2	$(2, 3)$	3	$(3, \infty)$
Test Point	-5	-4	0	2	2.5	3	4
Sign of $x + 4$	$-$	0	$+$	$+$	$+$	$+$	$+$
Sign of $x - 3$	$-$	$-$	$-$	$-$	$-$	0	$+$
Sign of $x - 2$	$-$	$-$	$-$	0	$+$	$+$	$+$
Sign of quotient	$-$	0	$+$	Undef.	$-$	0	$+$
Conclusion	Not Included	Included	Included	Not Included	Not Included	Included	Included

9.

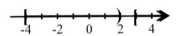

Section 5.7 Warm-Up

5.

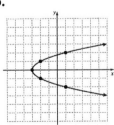

6.

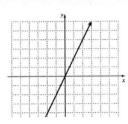

Warm Up 6.6

5b.

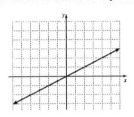

Guided Practice 6.6

5b.

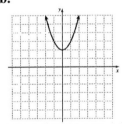

6b.

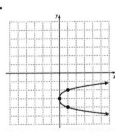

Do the Math 6.6

9.

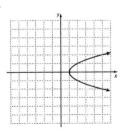

10.

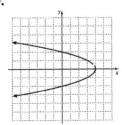

11.

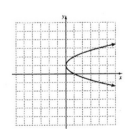

12.

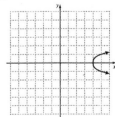

Sullivan/Struve, *Intermediate Algebra,* 3e
Copyright © 2014 Pearson Education, Inc

Warm Up 7.4

1.

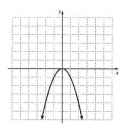

2.

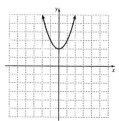

Do the Math 7.4

1.

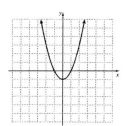

2.

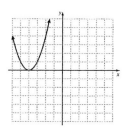

3.

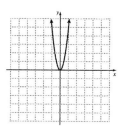

4.

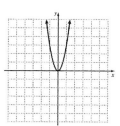

5.

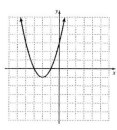

6.

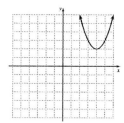

Guided Practice 7.5

3k.

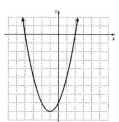

Do the Math 7.5

4.

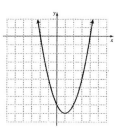

5.

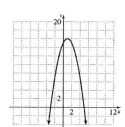

6.

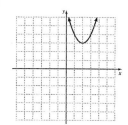

7.

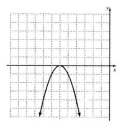

Sullivan/Struve, *Intermediate Algebra,* 3e
Copyright © 2014 Pearson Education, Inc.

GA-7

Do the Math 7.5

8.

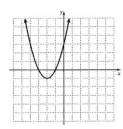

9.

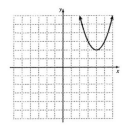

Guided Practice 7.6

3d.

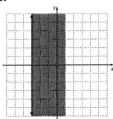

5d.

Interval	$(-\infty, -7)$	$x = -7$	$(-7, 5)$	$x = 5$	$(5, \infty)$
Test Point	-8	-7	0	5	6
Sign of $x + 7$	$-$	0	$+$	$+$	$+$
Sign of $x - 5$	$-$	$-$	$-$	0	$+$
Sign of product	$+$	0	$-$	0	$+$
Conclusion	Included	Not Included	Not Included	Not Included	Included

Do the Math 8.2

3.

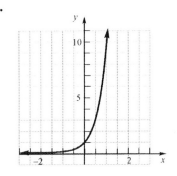

4.

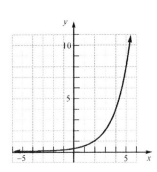

5.

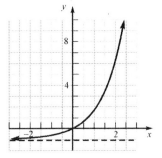

Guided Practice 8.3

7.

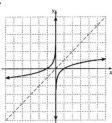

Do the Math 9.1

13a.

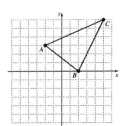

Sullivan/Struve, *Intermediate Algebra*, 3e
Copyright © 2014 Pearson Education, Inc

Guided Practice 9.2

4c.

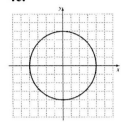

6d.

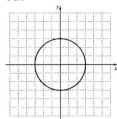

Do the Math 9.2

1.

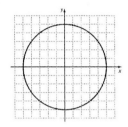

2.

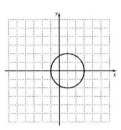

3.

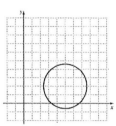

4.

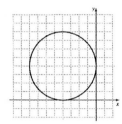

5.

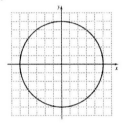

6.

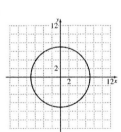

7.

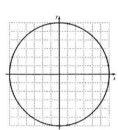

8.

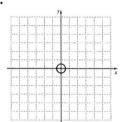

Guided Practice 9.3

5g.

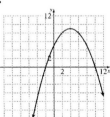

Do the Math 9.3

6.

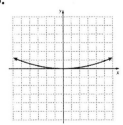

7.

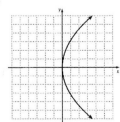

8.

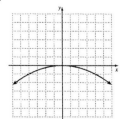

9.

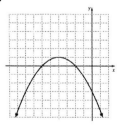

Do the Math 9.3

10.

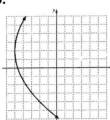

11.

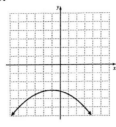

12.

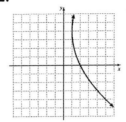

13.

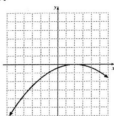

Guided Practice 9.4

3a.

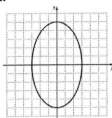

3b.

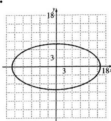

Do the Math 9.4

1.

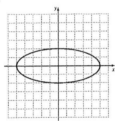

2.

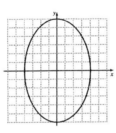

3.

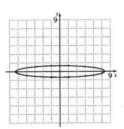

4.

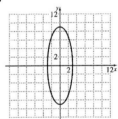

9.

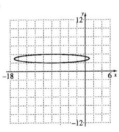

10.

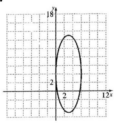

Warm Up 9.5

5.

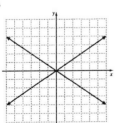

Guided Practice 9.5

3.

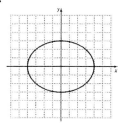

Do the Math 9.5

1.

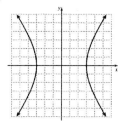

2.

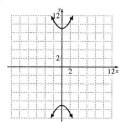

3.

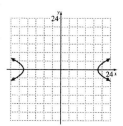

4.

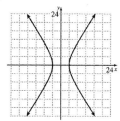

Guided Practice 9.6

1c.

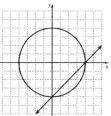

2c.

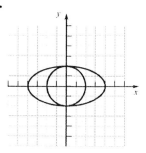